METHODS IN MOLECULAR BIOLOGY

Series Editor
**John M. Walker
School of Life and Medical Sciences
University of Hertfordshire
Hatfield, Hertfordshire, UK**

For further volumes:
http://www.springer.com/series/7651

For over 35 years, biological scientists have come to rely on the research protocols and methodologies in the critically acclaimed *Methods in Molecular Biology* series. The series was the first to introduce the step-by-step protocols approach that has become the standard in all biomedical protocol publishing. Each protocol is provided in readily-reproducible step-by-step fashion, opening with an introductory overview, a list of the materials and reagents needed to complete the experiment, and followed by a detailed procedure that is supported with a helpful notes section offering tips and tricks of the trade as well as troubleshooting advice. These hallmark features were introduced by series editor Dr. John Walker and constitute the key ingredient in each and every volume of the *Methods in Molecular Biology* series. Tested and trusted, comprehensive and reliable, all protocols from the series are indexed in PubMed.

Cell Cycle Checkpoints

Methods and Protocols

Edited by

James J. Manfredi

Department of Oncological Sciences, Icahn School of Medicine at Mount Sinai, New York, NY, USA

Humana Press

Editor
James J. Manfredi
Department of Oncological Sciences
Icahn School of Medicine at Mount Sinai
New York, NY, USA

ISSN 1064-3745   ISSN 1940-6029 (electronic)
Methods in Molecular Biology
ISBN 978-1-0716-1216-3   ISBN 978-1-0716-1217-0 (eBook)
https://doi.org/10.1007/978-1-0716-1217-0

© The Editor(s) (if applicable) and The Author(s) 2021
This work is subject to copyright. All rights are reserved by the Publisher, whether the whole or part of the material is concerned, specifically the rights of translation, reprinting, reuse of illustrations, recitation, broadcasting, reproduction on microfilms or in any other physical way, and transmission or information storage and retrieval, electronic adaptation, computer software, or by similar or dissimilar methodology now known or hereafter developed.
The use of general descriptive names, registered names, trademarks, service marks, etc. in this publication does not imply, even in the absence of a specific statement, that such names are exempt from the relevant protective laws and regulations and therefore free for general use.
The publisher, the authors, and the editors are safe to assume that the advice and information in this book are believed to be true and accurate at the date of publication. Neither the publisher nor the authors or the editors give a warranty, expressed or implied, with respect to the material contained herein or for any errors or omissions that may have been made. The publisher remains neutral with regard to jurisdictional claims in published maps and institutional affiliations.

This Humana imprint is published by the registered company Springer Science+Business Media, LLC, part of Springer Nature.
The registered company address is: 1 New York Plaza, New York, NY 10004, U.S.A.

Preface

The study of cell cycle checkpoints is becoming widespread, not only among cell and molecular biologists who have interests in the basic science of cell proliferation, but those interested in its implications for human disease, most notably, cancer. Targeting cell cycle proves to be an effective means for directed cancer therapy. Analysis of cell cycle checkpoints also serves as a means for prognosis and disease detection. It is hoped that this volume will serve a broad audience of scientists. Given the recent advances in the field, this also provides timely and much needed up-to-date information.

New York, NY, USA *James J. Manfredi*

Contents

Preface v
Contributors ix

1 Determination of CHK1 Cellular Localization by Immunofluorescence Microscopy 1
Yu-Che Cheng and Sheau-Yann Shieh

2 Detection of Post-translationally Modified p53 by Western Blotting 7
Anna Estevan Barber and David W. Meek

3 Global Analyses to Identify Direct Transcriptional Targets of p53 19
Matthew D. Galbraith, Zdenek Andrysik, Kelly D. Sullivan, and Joaquín M. Espinosa

4 Analysis of Replication Dynamics Using the Single-Molecule DNA Fiber Spreading Assay 57
Stephanie Biber and Lisa Wiesmüller

5 Measuring Translation Efficiency by RNA Immunoprecipitation of Translation Initiation Factors 73
Chris Lucchesi, Shakur Mohibi, and Xinbin Chen

6 DNA Affinity Purification: A Pulldown Assay for Identifying and Analyzing Proteins Binding to Nucleic Acids 81
Gerd A. Müller and Kurt Engeland

7 A Novel Strategy to Track Lysine-48 Ubiquitination by Fluorescence Resonance Energy Transfer 91
Kenneth Wu and Zhen-Qiang Pan

8 DNA Damage Response in *Xenopus laevis* Cell-Free Extracts 103
Tomas Aparicio Casado and Jean Gautier

9 Mammalian Cell Fusion Assays for the Study of Cell Cycle Progression by Functional Complementation 145
Jongkuen Lee and David Dominguez-Sola

10 Knockdown of Target Genes by siRNA In Vitro 159
Songhee Back and James J. Manfredi

11 Assaying Cell Cycle Status Using Flow Cytometry 165
Ramy Rahmé

12 Using TUNEL Assay to Quantitate p53-Induced Apoptosis in Mouse Tissues 181
Lois Resnick-Silverman

13 Generation and Analysis of dsDNA Breaks for Checkpoint and Repair Studies in Fission Yeast 191
Rohana Ramalingam and Matthew J. O'Connell

14	Calreticulin Exposure in Mitotic Catastrophe................................	207
	Lucillia Bezu, Oliver Kepp, and Guido Kroemer	
15	Quantification of eIF2α Phosphorylation Associated with Mitotic Catastrophe by Immunofluorescence Microscopy...........................	217
	Juliette Humeau, Lucillia Bezu, Oliver Kepp, Laura Senovilla, Peng Liu, and Guido Kroemer	
16	Clonogenic Assays to Detect Cell Fate in Mitotic Catastrophe.................	227
	José Manuel Bravo-San Pedro, Oliver Kepp, Allan Sauvat, Santiago Rello-Varona, Guido Kroemer, and Laura Senovilla	
Index...		*241*

Contributors

ZDENEK ANDRYSIK • *Department of Pharmacology, University of Colorado Anschutz Medical Campus, Aurora, CO, USA; Linda Crnic Institute for Down Syndrome, School of Medicine, University of Colorado Anschutz Medical Campus, Aurora, CO, USA*

TOMAS APARICIO CASADO • *Institute for Cancer Genetics, College of Physicians and Surgeons, Columbia University, New York, NY, USA*

SONGHEE BACK • *Department of Oncological Sciences and Graduate School of Biomedical Sciences, Icahn School of Medicine at Mount Sinai, New York, NY, USA*

ANNA ESTEVAN BARBER • *MRC Protein Phosphorylation Unit, University of Dundee, Dundee, Scotland, UK*

LUCILLIA BEZU • *Cell Biology and Metabomics Platforms, Gustaver Roussy Cancer Campus, Villejuif, France; Centre de Recherche des Cordeliers, Université de Paris, Sorbonne Université, Inserm U1138, Institut Universitaire de France, Paris, France*

STEPHANIE BIBER • *Department of Obstetrics and Gynecology, Ulm University, Ulm, Germany*

JOSÉ MANUEL BRAVO-SAN PEDRO • *Centre de Recherche des Cordeliers, Sorbonne Université, Inserm, Université de Paris, Equipe 11 Labellisée par la Ligue Contre le Cancer, Paris, France; Metabolomics and Cell Biology Platforms, Gustave Roussy Comprehensive Cancer Institute, Villejuif, France; Departamento de Fisiología, Facultad de Medicina, Universidad Complutense de Madrid, Madrid, Spain*

XINBIN CHEN • *Comparative Oncology Laboratory, University of California at Davis, Davis, CA, USA*

YU-CHE CHENG • *Institute of Biomedical Sciences, Academia Sinica, Taipei, Taiwan*

DAVID DOMINGUEZ-SOLA • *Department of Oncological Sciences, Icahn School of Medicine at Mount Sinai, New York, NY, USA; Tisch Cancer Institute, Icahn School of Medicine at Mount Sinai, New York, NY, USA; Precision Immunology Institute, Icahn School of Medicine at Mount Sinai, New York, NY, USA; Graduate School of Biomedical Sciences, Icahn School of Medicine at Mount Sinai, New York, NY, USA*

KURT ENGELAND • *Molecular Oncology, Faculty of Medicine, Leipzig University, Leipzig, Germany*

JOAQUÍN M. ESPINOSA • *Department of Pharmacology, University of Colorado Anschutz Medical Campus, Aurora, CO, USA; Linda Crnic Institute for Down Syndrome, School of Medicine, University of Colorado Anschutz Medical Campus, Aurora, CO, USA; Department of Molecular, Cellular and Developmental Biology, University of Colorado Boulder, Boulder, CO, USA*

MATTHEW D. GALBRAITH • *Department of Pharmacology, University of Colorado Anschutz Medical Campus, Aurora, CO, USA; Linda Crnic Institute for Down Syndrome, School of Medicine, University of Colorado Anschutz Medical Campus, Aurora, CO, USA*

JEAN GAUTIER • *Institute for Cancer Genetics, College of Physicians and Surgeons, Columbia University, New York, NY, USA; Department of Genetics and Development, Columbia University School of Medicine, New York, NY, USA*

JULIETTE HUMEAU • *Centre de Recherche des Cordeliers, Université de Paris, Sorbonne Université, Inserm U1138, Institut Universitaire de France, Paris, France*

OLIVER KEPP • *Cell Biology and Metabomics Platforms, Gustaver Roussy Cancer Campus, Villejuif, France; Centre de Recherche des Cordeliers, Université de Paris, Sorbonne Université, Inserm U1138, Institut Universitaire de France, Paris, France*

GUIDO KROEMER • *Cell Biology and Metabomics Platforms, Gustaver Roussy Cancer Campus, Villejuif, France; Centre de Recherche des Cordeliers, Université de Paris, Sorbonne Université, Inserm U1138, Institut Universitaire de France, Paris, France*

JONGKUEN LEE • *Graduate School of Biomedical Sciences, Icahn School of Medicine at Mount Sinai, New York, NY, USA*

PENG LIU • *Centre de Recherche des Cordeliers, Université de Paris, Sorbonne Université, Inserm U1138, Institut Universitaire de France, Paris, France*

CHRIS LUCCHESI • *Comparative Oncology Laboratory, University of California at Davis, Davis, CA, USA*

JAMES J. MANFREDI • *Department of Oncological Sciences and Graduate School of Biomedical Sciences, Icahn School of Medicine at Mount Sinai, New York, NY, USA*

DAVID W. MEEK • *MRC Protein Phosphorylation Unit, University of Dundee, Dundee, Scotland, UK*

SHAKUR MOHIBI • *Comparative Oncology Laboratory, University of California at Davis, Davis, CA, USA*

GERD A. MÜLLER • *Molecular Oncology, Faculty of Medicine, Leipzig University, Leipzig, Germany; Department of Chemistry and Biochemistry, University of California, Santa Cruz, Santa Cruz, CA, USA*

MATTHEW J. O'CONNELL • *Department of Oncological Sciences, Icahn School of Medicine at Mount Sinai, New York, NY, USA*

ZHEN-QIANG PAN • *Department of Oncological Sciences, Icahn School of Medicine at Mount Sinai, New York, NY, USA*

RAMY RAHMÉ • *Department of Oncological Sciences, Icahn School of Medicine at Mount Sinai, New York, NY, USA*

ROHANA RAMALINGAM • *Department of Oncological Sciences, Icahn School of Medicine at Mount Sinai, New York, NY, USA*

SANTIAGO RELLO-VARONA • *Centre de Recherche des Cordeliers, Sorbonne Université, Inserm, Université de Paris, Equipe 11 Labellisée par la Ligue Contre le Cancer, Paris, France; Metabolomics and Cell Biology Platforms, Gustave Roussy Comprehensive Cancer Institute, Villejuif, France; Cell Biology Department, School of Biological Sciences, Universidad Complutense de Madrid—UCM, Madrid, Spain; Hospital La Paz Institute for Health Research—IdiPAZ, Madrid, Spain*

LOIS RESNICK-SILVERMAN • *Department of Oncological Sciences, Icahn School of Medicine at Mount Sinai, New York, NY, USA*

ALLAN SAUVAT • *Centre de Recherche des Cordeliers, Sorbonne Université, Inserm, Université de Paris, Equipe 11 Labellisée par la Ligue Contre le Cancer, Paris, France; Metabolomics and Cell Biology Platforms, Gustave Roussy Comprehensive Cancer Institute, Villejuif, France*

LAURA SENOVILLA • *Centre de Recherche des Cordeliers, Sorbonne Université, Inserm, Université de Paris, Equipe 11 Labellisée par la Ligue Contre le Cancer, Paris, France; Metabolomics and Cell Biology Platforms, Gustave Roussy Comprehensive Cancer Institute, Villejuif, France*

SHEAU-YANN SHIEH • *Institute of Biomedical Sciences, Academia Sinica, Taipei, Taiwan*

KELLY D. SULLIVAN • *Department of Pediatrics, School of Medicine, University of Colorado Anschutz Medical Campus, Aurora, CO, USA; Linda Crnic Institute for Down Syndrome, School of Medicine, University of Colorado Anschutz Medical Campus, Aurora, CO, USA*
LISA WIESMÜLLER • *Department of Obstetrics and Gynecology, Ulm University, Ulm, Germany*
KENNETH WU • *Department of Oncological Sciences, Icahn School of Medicine at Mount Sinai, New York, NY, USA*

Chapter 1

Determination of CHK1 Cellular Localization by Immunofluorescence Microscopy

Yu-Che Cheng and Sheau-Yann Shieh

Abstract

Many proteins involved in the DNA damage pathway shuttle between the cytoplasm and nucleus, and their localizations are important for functions. In that regard, immunofluorescence microscopy has been widely used to delineate the temporal and spatial regulation of proteins. Here, we describe an unconventional method for studying the cellular localization of CHK1, a cell cycle checkpoint kinase that undergoes shuttling from the cytoplasm to the nucleus in response to genotoxic stress. In this study, we included an acid extraction step to better reveal the nuclear localization of CHK1.

Key words CHK1, DNA damage response, HCl extraction, Immunofluorescence microscopy, Nuclear-cytoplasmic shuttling

1 Introduction

The DNA damage response (DDR) pathway, which involves both cell cycle control and DNA repair, is critical for cell survival and maintenance of genome stability. ATR-CHK1 and ATM-CHK2 are two main signaling axes in the DDR network [1]. The ATM-CHK2 responds to double-strand breaks, whereas the ATR-CHK1 pathway responds primarily to single-strand DNA generated at the stalled replication fork or in the double-strand break repair process [2, 3]. CHK1 is a key effector kinase in the DDR and is activated through phosphorylation at Ser317 and Ser345 by the ATR kinase upon UV exposure or replication stress [4]. CHK1 phosphorylates many downstream substrates and is required for maintenance of cell cycle checkpoint as well as cell survival [5, 6]. For example, CHK1 maintains cell survival after treatment with hydroxyurea (HU), a ribonucleotide reductase inhibitor [7]. In addition, CHK1 also functions in the spindle assembly checkpoint to ensure proper kinetochore attachment and alignment of condensed chromosomes. CHK1 phosphorylates its target Aurora B and promotes its kinase activity to enhance BubR1 kinetochore localization after paclitaxel treatment [8].

In addition to ATR-mediated CHK1 phosphorylation, CPT treatment promotes CHK1 to undergo CUL1- and CUL4-dependent polyubiquitination and degradation [9]. DNA damage-induced CHK1 phosphorylation also promotes its cytoplasmic translocation and subsequent degradation by the E3 ligase Fbx6 [10]. In our previous studies, we have demonstrated that BTG3 promotes CHK1 K63-linked polyubiquitination, which contributes to its nuclear localization, chromatin association, and activation to maintain cell survival and the spindle assembly checkpoint [11]. Based on these studies, we know that the cellular localization of CHK1 is important for its functions and regulation. It is therefore necessary to develop an effective method to reproducibly demonstrate the distribution of CHK1 in cells. In this report, we describe in detail the materials and methods for the determination of CHK1 cellular localization by immunofluorescence. This protocol can also be applied to other DDR proteins shuttling between the cytoplasm and nucleus.

2 Materials

1. HCT116 cells.
2. RPMI medium.
3. Fetal bovine serum (FBS).
4. 10,000 U/mL penicillin and 10 mg/mL streptomycin.
5. Cover glasses.
6. Microscope slides.
7. Plasmid DNA.
8. Transfection reagent.
9. UV Crosslinker.
10. Phosphate buffered saline (PBS): 137 mM NaCl, 2.7 mM KCl, 4.3 mM Na_2HPO_4, and 1.47 mM KH_2PO_4. Adjust to a final pH of 7.4.
11. Paraformaldehyde (PFA): 4% solution in PBS, made fresh.
12. 2 N HCl in H_2O, made fresh.
13. 0.1 M Sodium borate pH 8.5.
14. Blocking solution: 5% bovine serum albumin (BSA) in PBS.
15. Antibody dilution buffer: 1% BSA in PBS.
16. Mouse anti-CHK1 antibody (sc8408) from Santa Cruz Biotechnology.
17. Mouse anti-HA antibody from LTK BioLaboratories.

18. DyLight488-conjugated goat anti-mouse IgG antibody (115-485-006) from Jackson ImmunoResearch.
19. 4′,6-Diamidino-2-phenylindole (DAPI): Prepare 5 mg/mL stock in H$_2$O and store at −20 °C. Dilute to 5 μg/mL in working solution.
20. 1% 1,4-Diazabicyclo[2.2.2]octane (DABCO) solution: Dissolve DABCO power in 90% glycerol and store at −20 °C.
21. Confocal microscope.
22. Software for quantification.

3 Methods

3.1 Cell Culture and Treatment

1. Grow HCT116 cells in RPMI medium containing 10% FBS and 1% penicillin/streptomycin.
2. Seed HCT116 cells on coverslips 1 day before transfection (*see* **Note 1**).
3. Transfect cells with the indicated plasmids using the appropriate transfection reagent (*see* **Note 2**).
4. Treat cells with UVC using the UV Crosslinker the next day after transfection (*see* **Note 3**).

3.2 Immunofluorescence Microscopy

1. Fix cells in 2% PFA at 4 °C overnight (*see* **Note 4**).
2. Wash the coverslips once with PBS.
3. Permeabilize cells in 0.5% Triton X-100/PBS for 10 min at room temperature.
4. Wash the coverslips once with PBS.
5. Incubate the coverslips in 2 N HCl on ice for 30 min or go directly to **step 8** (*see* **Note 5**).
6. Discard the HCl solution and neutralize with 1 M sodium borate.
7. Wash the coverslips once with PBS.
8. Block in 5% BSA in PBS at room temperature for 30 min.
9. Wash the coverslips once with PBS.
10. Incubate the coverslips with mouse anti-CHK1 or anti-HA antibody at 4 °C overnight in a humidified multi-well culture dish (*see* **Notes 6** and **7**).
11. Wash the coverslips once with PBS.
12. Incubate with DyLight488-conjugated goat anti-mouse IgG antibody and DAPI at room temperature for 1 h (*see* **Note 8**).
13. Wash the coverslips once with PBS.
14. Mount the coverslips on microscope slides in DABCO solution (*see* **Note 9**).

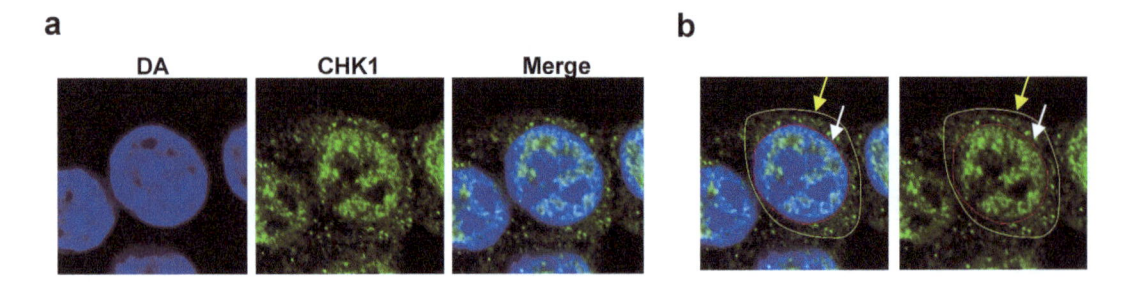

Fig. 1 Determination of the nucleus-to-cytoplasm (N/C) ratio of CHK1. (**a**) HCT116 cells were treated with UV and the localization of CHK1 was determined by confocal microscopy. (**b**) The total (marked by yellow arrows) and the nuclear (marked by white arrows) regions were gated and quantified using confocal microscopy software (ZEN2011). (**c**) The total intensity of the gated area was exported as Excel files and used to derive the N/C ratio

	Total (N + C)	Nucleus	Cytoplasm	N/C ratio
Mean Intensity	42.61601	52.79236		
Area [μm x μm]	218.92	131.74		
Total Intensity	9329.49691	6954.86551	2374.6314	2.92881897

15. Seal the coverslips with transparent nail polish.
16. Acquire images by confocal microscopy (*see* **Note 10**).
17. Quantify the intensity of the acquired images by software and calculate the N/C ratio (*see* Fig. 1).

4 Notes

1. The seeding density may depend upon the cell line and transfection reagent used. To achieve optimal results, run preliminary experiments with several seeding densities. For transfecting HCT116 cells with plasmid, we typically seed 2×10^5 cells per 35 mm dish using Lipofectamine 2000 (Invitrogen) as a transfection reagent. The cells are collected at 60–70% confluency.

2. Protein overexpression may alter properties such as cellular localization. Our results show that overexpression of CHK1 will enhance its nuclear localization (*see* Fig. 2).

3. The UV dosage and recovery time should be tested for in the cell line being used to find the optimal conditions for treatment. To ensure equal transfection efficiency, cells are transfected in the same dish and then reseeded onto new dishes containing coverslips to obtain a temporal and dose response.

4. The PFA solution should be freshly prepared and dissolved in PBS. The solution may appear cloudy initially. Vortexing as well

Immunofluorescence Detection of CHK1   5

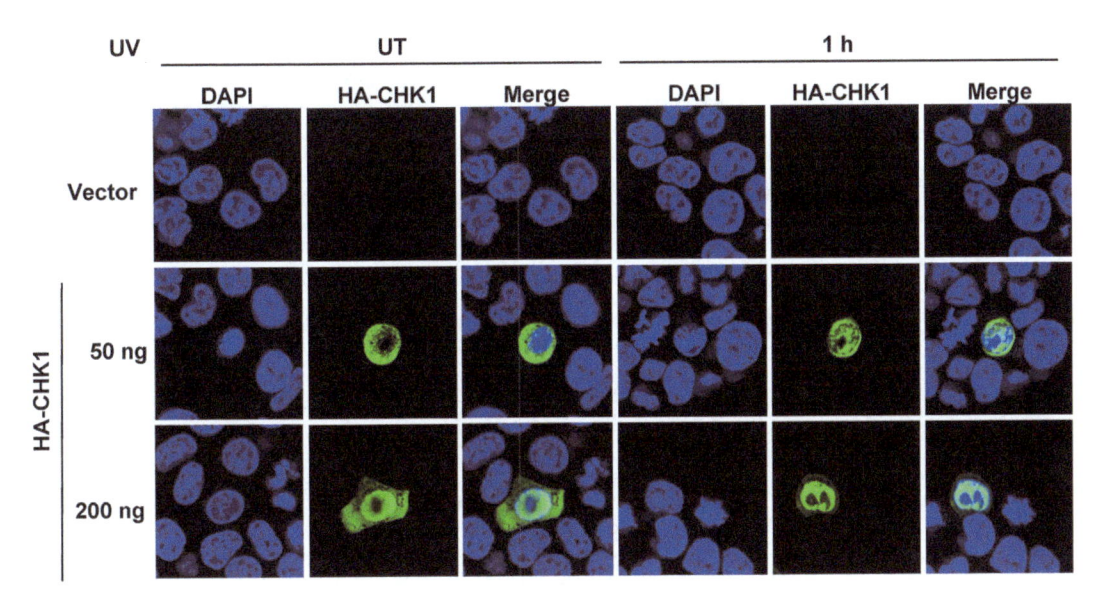

Fig. 2 Overexpression of CHK1 promotes its nuclear localization. HCT116 cells were transfected with the indicated amounts of the plasmid expressing HA-CHK1. Cells were treated with UVC (20 J/m^2) or untreated (UT), and the cellular localization of HA-CHK1 was determined by confocal microscopy following staining with the anti-HA antibody

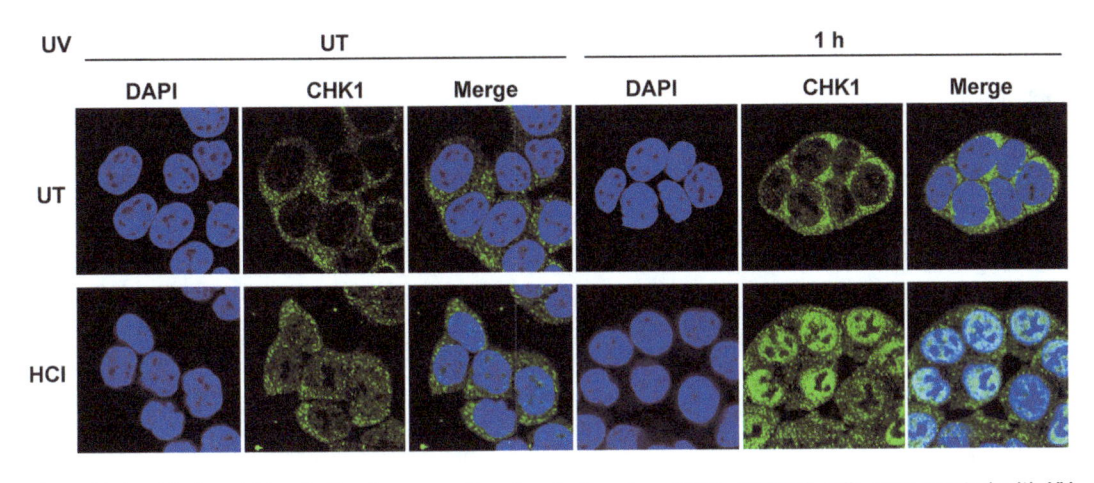

Fig. 3 Hydrochloric acid incubation improves the signal of nuclear CHK1. HCT116 cells were treated with UV and fixed in 2% paraformaldehyde overnight. After permeabilization, cells were incubated with or without 2 N HCl and probed with the CHK1 antibody. The cellular localization of CHK1 was determined by confocal microscopy

as heating the solution to 65 °C can help promote dissolution. After the UV treatment, we add 1 mL of 4% PFA to the dish containing 1 mL of medium for a final concentration of 2% PFA.

5. The 2 N HCl was diluted in H$_2$O, not PBS. The HCl treatment greatly enhances the detection of nucleus-localized CHK1, especially after the UV treatment (*see* Fig. 3).

6. We prepare primary antibody (mouse anti-CHK1) with 1% BSA in PBS. The dilution factor should be tested for different batches of antibody. To confirm the specificity, incubation without the primary antibody can be performed as a negative control.

7. We routinely use 6-well or 12-well culture dish (depending on the size of coverslips) for antibody incubation. First, add a drop of antibody solution (30–50 μL) to the well, and then transfer the coverslips to the well, topside down, so that the fixed cells will be in full contact with the antibody solution.

8. The DyLight488-conjugated secondary antibody gives a stronger signal than the FITC-conjugated secondary antibody; therefore the dilution factor could be increased to 1:500–1000.

9. Care should be taken to avoid bubbles when mounting the coverslips onto the slides.

10. The slides can be stored at 4 °C before confocal microscopy.

Acknowledgments

This work was supported by funding from the Academia Sinica and the Ministry of Science and Technology Taiwan.

References

1. Ciccia A, Elledge SJ (2010) The DNA damage response: making it safe to play with knives. Mol Cell 40(2):179–204
2. Abraham RT (2001) Cell cycle checkpoint signaling through the ATM and ATR kinases. Genes Dev 15(17):2177–2196
3. Myers JS, Cortez D (2006) Rapid activation of ATR by ionizing radiation requires ATM and Mre11. J Biol Chem 281(14):9346–9350
4. Zhao H, Piwnica-Worms H (2001) ATR-mediated checkpoint pathways regulate phosphorylation and activation of human Chk1. Mol Cell Biol 21(13):4129–4139
5. Latif C, den Elzen NR, O'Connell MJ (2004) DNA damage checkpoint maintenance through sustained Chk1 activity. J Cell Sci 117. (Pt 16:3489–3498
6. Meuth M (2010) Chk1 suppressed cell death. Cell Div 5:21
7. Rodriguez R, Meuth M (2006) Chk1 and p21 cooperate to prevent apoptosis during DNA replication fork stress. Mol Biol Cell 17(1):402–412
8. Zachos G, Black EJ, Walker M, Scott MT, Vagnarelli P, Earnshaw WC, Gillespie DA (2007) Chk1 is required for spindle checkpoint function. Dev Cell 12(2):247–260
9. Zhang YW, Otterness DM, Chiang GG, Xie W, Liu YC, Mercurio F, Abraham RT (2005) Genotoxic stress targets human Chk1 for degradation by the ubiquitin-proteasome pathway. Mol Cell 19(5):607–618
10. Zhang YW, Brognard J, Coughlin C, You Z, Dolled-Filhart M, Aslanian A, Manning G, Abraham RT, Hunter T (2009) The F box protein Fbx6 regulates Chk1 stability and cellular sensitivity to replication stress. Mol Cell 35(4):442–453
11. Cheng YC, Lin TY, Shieh SY (2013) Candidate tumor suppressor BTG3 maintains genomic stability by promoting Lys63-linked ubiquitination and activation of the checkpoint kinase CHK1. Proc Natl Acad Sci U S A 110(15):5993–5998

Chapter 2

Detection of Post-translationally Modified p53 by Western Blotting

Anna Estevan Barber and David W. Meek

Abstract

The p53 tumor suppressor has a central role in many key cellular processes including the DNA damage response, aging, stem cell differentiation, and fertility. p53 undergoes extensive regulatory post-translational modification through events such as phosphorylation, acetylation, methylation, and ubiquitylation. Here, we describe western blotting-based methodology for the detection and relative quantification of individual phosphorylation events in p53. While we focus on well-established N-terminal modifications for the purpose of illustration, this approach can be used to investigate other post-translational modifications of the protein, drawing upon a broad range of commercially available modification-specific antibodies.

Key words Phosphorylation, DNA-damage response, p53, Post-translational modifications, Phospho-specific antibodies, Western blot

1 Introduction

The p53 tumor suppressor is a potent transcription factor [1, 2], which is subject to a wide range of post-translational modifications (PTMs) that tightly control its activity and contribute to regulating its stability. Many of these modifications occur in response to various forms of genotoxic or nongenotoxic stresses [3–5].

Phosphorylation of the N-terminus of p53 occurs rapidly in response to DNA damage (for a detailed review, see Meek [6]). Serine 15 phosphorylation, catalyzed mainly by the ATM and ATR protein kinases, is a key modification event that regulates p53 function and stability. Serine 15 phosphorylation is also a nucleation event that is required for sequential threonine 18 phosphorylation by CK1. ATM also activates Chk2, which in turn phosphorylates serine 20 in p53. These phosphorylation events stimulate the interaction of p53, in a cooperative manner, with transcriptional co-activators such as p300 and CBP, leading to the acetylation of lysines in the p53 C-terminus, principally K320,

K372, K373, K381, K382, and K386. In addition, serine 20 phosphorylation contributes to uncoupling p53 from its negative regulator MDM2, especially in cooperation with serine 15 and threonine 18 phosphorylation. Collectively these events result in p53 activation and contribute significantly, along with MDM2 inactivation, to increased p53 stability [6]. Upon activation, p53 controls the expression of genes involved in a range of cellular responses including cell cycle arrest and apoptosis [1, 2]. p53 also has important roles in other key biological processes including metabolism, aging, stem cell differentiation, and fertility, but the role of regulatory PTMs in these processes is poorly understood.

p53 function is essential to mediate innate tumor suppression as it eliminates cells with increased proliferative potential and genetic abnormalities. Consistent with such a role, p53 inactivating mutations are found in about 50% of human cancers [7]. The central DNA-binding domain holds approximately 95% of the p53 mutations encountered in tumors and the vast majority are of the missense type. Among p53 mutations are six "hot spot" residues that are altered with high frequency: R175, G245, R248, R249, R273, and R282 [8]. p53 missense mutations lead to the expression of full-length mutant proteins which are structurally altered and have a prolonged half-life as compared with the wild-type protein. The retention of mutant versions of p53 suggests that these proteins may provide a selective oncogenic advantage. In fact, in addition to the loss of the wild-type tumor suppression function, mutant p53 has been widely reported to exert a dominant negative effect towards wild-type p53 and/or display novel oncogenic activities that are independent of the wild-type p53 function [8, 9]. There is evidence that mutant p53 function can be activated in vivo in response to genotoxic drugs [10]. However, the possibility that post-translational modifications can impact mechanistically on the gain of function properties of mutant p53 is still uncertain and is the subject of many investigations [11].

In these contexts, detection of PTMs on p53 has become essential in order to monitor the activation of both wild-type and mutant p53 proteins under specific conditions of interest, such as when investigating the response to DNA damage-based chemotherapeutic treatments. Western blotting has been the main research tool used to study post-translationally modified p53 as it provides sensitivity, specificity in studying individual sites of modification, a means of quantifying changes in modification status, and ease of application. Ensuring that the antibodies used recognize only modified p53 is therefore crucial for accurate interpretation of results. Suppliers normally verify the specificities of antibodies raised against specific PTMs on p53 through a decrease or increase of the signal upon normal or genotoxic conditions, respectively. However, further validation is encouraged to ensure that optimal conditions are achieved to minimize the possibilities of cross-reaction with other modified sites in p53 or, indeed, with unmodified p53 itself. Additionally, we have used phospho-mimic (serine/aspartate substitution) and non-phosphorylatable (serine/alanine substitution) mutants to confirm specificity and optimize the conditions for detecting p53 only when it is phosphorylated at serine 15, serine 20 and/or acetylated at lysine 382. Here, for the purpose of demonstration, we have used R175H p53-expressing cancer cells as a model.

2 Materials

2.1 Standard SDS-PAGE

1. For all p53 western blotting detection procedures, we routinely carry out standard SDS-PAGE, using the ATTO AE-6500 Dual Mini Slab electrophoresis system (*see* **Note 1**).

2.2 Western Blot Transfer

1. Whatman filter paper (Fisher Scientific): cut filter sheets of 9 cm × 7 cm.

2. Hybond ECL nitrocellulose membrane (GE healthcare): cut membrane pieces of 9 cm × 6 cm.

3. Transfer buffer: 0.025 M Tris, 0.192 M Glycine, and 20% (v/v) Methanol. Prepare 1 l of 10× transfer buffer (mix 30 g of Tris and 144 g of glycine and make up to 1 l with dH$_2$O), and store the solution at room temperature. Before use, prepare 1× transfer buffer by mixing 100 ml of 10× transfer buffer, 200 ml of methanol, and 700 ml of dH$_2$O.

4. Wet transfer blotting system (Bio-Rad): We use the Mini Trans-Blot® Cell system.

5. Ponceau red protein staining buffer: 0.2% (w/v) Ponceau S and 5% (v/v) Glacial acetic acid.

6. TBS-T (Tris buffered saline-Tween20): 20 mM Tris, pH 7.6, 0.14 M NaCl, and 0.1% (v/v) Tween-20. Prepare 1 ml of 10× TBS by mixing 24.2 g Tris and 80 g NaCl into 1 ml graduated glass bottle. Add dH$_2$O to a volume of 900 ml and dissolve using a magnetic stir bar. Adjust pH to 7.6 pH using HCl. Top up to 1 l with dH$_2$O. Before use prepare 1× TBS-T by mixing 100 ml of 10× TBS, 890 ml of dH$_2$O, and 10 ml of 10% (v/v) Tween-20.

2.3 Western Blot Development

1. Blocking buffer for the membrane: 5% (w/v) Marvel dried skimmed milk in TBS-T (*see* **Note 2**).

2. Dilution buffer for the antibodies: 5% (w/v) Bovine serum albumin (BSA) (*see* **Note 3**) or Marvel dried skimmed milk in TBS-T as indicated in Table 1.

3. Antibodies: these are listed in Table 1 (*see* **Note 4**).

4. Clarity enhanced chemiluminescence solution (Bio-Rad).

Table 1
Description of the antibodies used

Primary antibodies	Dilution in blocking buffer	Amount of protein loaded (μg)	Supplier
Phospho-serine 20 p53 (produced in rabbit)	1:5000 in 5% (w/v) BSA	30	Abcam (ab157454)
Phospho-serine 15 p53 (produced in rabbit)	1:1000 in 5% (w/v) milk	15–20	Cell signaling (9284)
Acetyl-lysine 382 p53 (produced in rabbit)	1:700 in 5% (w/v) BSA	30	Cell signaling (2525)
Flag (produced in mouse)	1:1000 in 5% (w/v) milk	15	Sigma (F1804)
p53 (1801) (produced in mouse)	1:1000 in 5% (w/v) milk	15	Santa Cruz (SC-98)
Phospho-serine 139 histone H2AX (produced in rabbit)	1:1000 in 5% (w/v) milk	15–20	Cell signaling (2577)

Secondary antibodies	Dilution in blocking buffer	Supplier
Anti-mouse IgG	1:2000 in 5% (w/v) milk	Bio-Rad (1721011)
Anti-rabbit IgG	1:2000 in 5% (w/v) milk	Bio-Rad (1706515)

3 Methods

3.1 Protein Extraction and Sample Preparation for SDS-PAGE

The protocol describes the procedure for extracting proteins from adherent cells that were previously seeded in a 6-well plate and harvested at 70–80% confluence (*see* **Note 5**).

1. Place the plate on ice and wash twice with 1 ml/well of PBS.

2. Add 200 μl of 2× SDS protein extraction buffer (0.125 M Tris–HCl (pH 6.8), 20% (v/v) Glycerol, 4% (w/v) Sodium dodecyl sulfate (SDS), 0.02% (w/v) Bromophenol blue, and no reducing agent) directly to the wells and use a cell scraper to collect the lysate. Pipette the solution into a microcentrifuge tube. Incubate for 15 min on ice. Vortex every 3–5 min (*see* **Note 6**). Sonicate the lysate for 20 s at 30% amplitude before centrifuging at 10,000 RCF for 8 min at room temperature.

3. Transfer the supernatant to a new tube. Determine the protein concentration using the DC Protein Assay (Bio-Rad), using

BSA as protein standard. Dilute the sample in 2× SDS protein extraction buffer to 1–2 μg/μl. Add DTT to a final concentration of 0.1 M and mix by vortexing (*see* **Note 7**).

4. Heat the samples for 5 min at 98 °C. Leave the samples at room temperature to cool down and centrifuge briefly to recover any water condensed at the top of the tubes. If the samples are going to be used immediately, keep the tubes on ice. Alternatively, they can be stored at −20 °C or −80 °C for later use.

3.2 SDS-PAGE (See Note 1)

For p53, we normally cast 8% resolving gels, using the ATTO AE-64011 mm Dual Mini Gel Cast. The gel size is 90 mm (W) × 80 mm (H) × 1 mm (D) (*see* **Note 8**). Ensure that pre-stained protein markers are included on every gel (*see* **Note 9**).

3.3 Western Blot Transfer

1. Separate the glass plates using a spatula. Remove the stacking gel with a cutter. Place the glass containing the gel in a container with enough transfer buffer to cover the gel. Carefully separate the gel from the glass.

2. Wet the sponges, filter papers, and membranes in transfer buffer. In a container with a small amount of transfer buffer, make a transfer sandwich on the black side of the transfer cassette as follows: sponge, 2 filter papers, gel, membrane, 2 filter papers, sponge (*see* **Note 10** and Fig. 1). Close the transfer cassette and move it to the transfer tank. Insert an ice pack and fill the tank with transfer buffer. Transfer for 1 h at 100 V (*see* **Note 11**).

3. Move the membrane into a plastic box and add sufficient ponceau red solution to cover the membrane. Wait for 1 min. Discard the solution and wash the membrane with water two or three times until the protein bands are well defined (*see* **Note 12**).

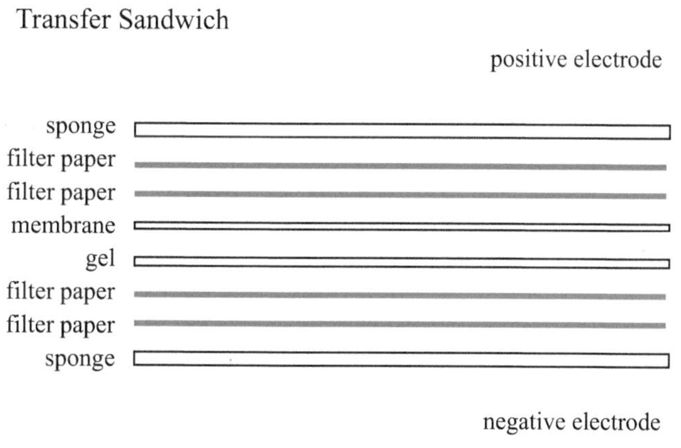

Transfer Sandwich

positive electrode

sponge
filter paper
filter paper
membrane
gel
filter paper
filter paper
sponge

negative electrode

Fig. 1 Schematic of the transfer sandwich. See the text for details (Subheading 3.3)

3.4 Western Blot Development

1. Place the membrane in a 50 ml tube containing 40 ml of blocking buffer. Place the tube on a roller for 1 h at room temperature to ensure continuous flow of the solution over the membrane.

2. Discard the blocking buffer. Rinse the membrane by adding 20–30 ml of TBS-T and placing the tube on the roller for 1 min. Discard the TBS-T and add 5 ml of the primary antibody solution prepared as indicated in Table 1. Place the tubes on the roller and incubate overnight at 4 °C (*see* **Note 13**).

3. Wash the membrane by adding approximately 30 ml of TBS-T and place the tube on the roller for 5 min at room temperature. Repeat this three times.

4. Add the secondary antibody, as indicated in Table 1, and incubate for 1 h on the roller at room temperature. Wash the membrane three times as in **step 3**.

5. Prepare the ECL (*E*nhanced *C*hemi-*L*uminescence: Bio-Rad Clarity™) solution, as instructed by the manufacturer. Approximately 1 ml of solution is needed per membrane. Move the membrane into a plastic transparent tray and add the solution using a 1000 μl pipette. Rock the tray briefly to ensure the solution covers the whole membrane.

6. Visualize the antibody-bound protein(s) on the membrane(s) by chemiluminescence, using The ChemiDoc™ MP imaging System (Bio-Rad) controlled by the Image Lab 4.1 software (*see* **Note 14**).

7. Examples showing the analysis of post-translational modification of endogenous mutant p53 in SKBR3 cells, and of mutant p53 proteins ectopically expressed in H1299 cells, are given in Figs. 2, 3, and 4, together with a brief interpretation of these data.

4 Notes

1. While the ATTO system has been our preferred system of choice, other researchers may prefer to use commercially available precast gel systems, which are also appropriate for the detection of post-translationally modified forms of p53. We normally prepare gels as follows. *Resolving gel:* 0.382 M Tris–HCl (pH 8.8), 0.1% SDS (w/v), 0.1% (w/v) APS, 0.1% (v/v) of N,N,N,N'-Tetramethylethylenediamine (TEMED), and 8% Acrylamide/Bis-Acrylamide. *Stacking gel:* 0.123 M Tris–HCl (6.8), 0.1% SDS (w/v), 0.1% (w/v) APS, 0.16% (v/v) TEMED, and 4.9% (w/v) Acrylamide/Bis-Acrylamide. *Running buffer:* 0.025 M Tris, 0.192 M Glycine, 0.1 (w/v) % SDS.

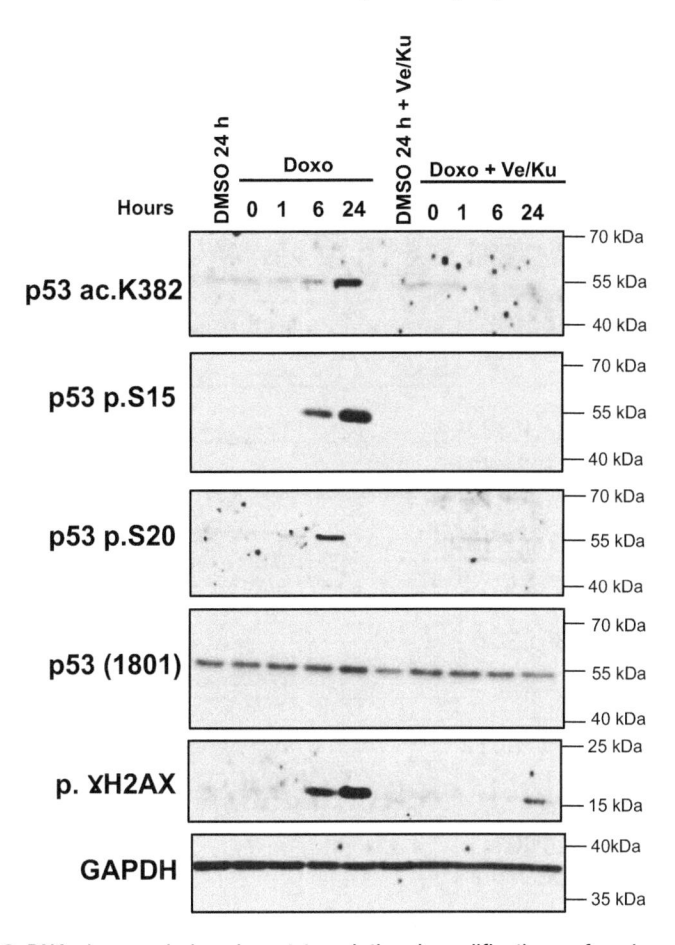

Fig. 2 DNA damage-induced post-translational modifications of endogenous R175H mutant p53 in SKBR3 cells. SKBR3 cells were treated with 0.03 μM doxorubicin or DMSO and harvested at the indicated time points. Where indicated, cells were pre-treated for 1 h with 7 μM Ku-55933 (ATM inhibitor, Ku) and 2.5 μM Ve-821 (ATR inhibitor, Ve). The inhibitors were maintained during the doxorubicin treatment. Western blotting was subsequently performed. Phosphorylation of serines 15 and 20 as well as acetylation of lysine 382 increased in a time-dependent manner upon doxorubicin treatment. Treatment with ATM and ATR kinase inhibitors reduced the levels of the three PTMs tested. This suggests that, similar to wt p53, ATM and ATR can mediate mutant p53 phosphorylation at serine 15 and 20 under genotoxic stress. p53 N-terminal phosphorylation events, mediated mainly by ATM and ATR, recruit the histone acetyl transferases (HAT) p300/CBP, which subsequently acetylate wt p53 at various C-terminal lysines. Consistent with this model, lysine 382 acetylation in mutant p53 decreased following ATM and ATR inhibition. Doxorubicin treatment also promoted mutant p53 accumulation. Total levels of mutant p53s were downregulated under ATM and ATR inhibition in both normal conditions and doxorubicin treatment. These data support the idea that PTMs play a role in promoting mutant p53 stability and suggest that wt p53 and mutant p53 may share upstream stabilization and activation pathways. Phosphorylated H2AX was used as a DNA damage marker

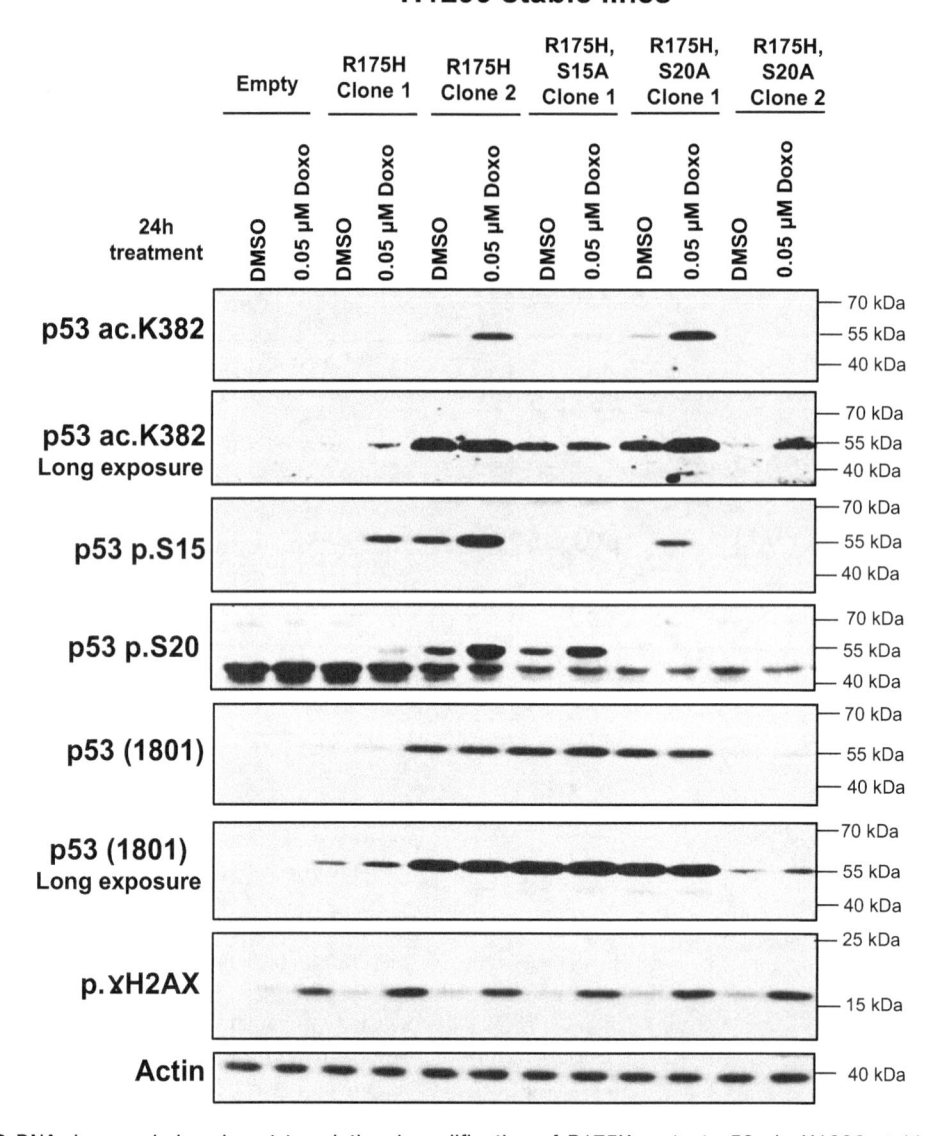

Fig. 3 DNA damage-induced post-translational modification of R175H mutant p53s in H1299 stable lines. H1299 stable lines were generated expressing flag-tagged R175H mutant p53 and non-phosphorylatable versions of R175H mutant, obtained by changing serine 15 or serine 20 to alanine. Clones expressing low and high levels of mutant p53 are shown as indicated. Cells were treated with 0.05 µM doxorubicin or DMSO and harvested after 24 h. Western blot was subsequently performed. The specificity of the phospho-serine 15 and phospho-serine 20 antibodies was verified by the absence of phosphorylation in the respective serine to alanine substitutions. Acetylation of lysine 382 was abolished in cells expressing R175H-S15A mutant p53 but not in R175H mutant p53, consistent with the idea that, as with wild-type p53, the histone acetyl-transferase (HAT) p300 is recruited to acetylated lysine 382, and supporting the specificity of the acetyl-lysine 382 antibody. Phosphorylated H2AX was used as a DNA damage marker

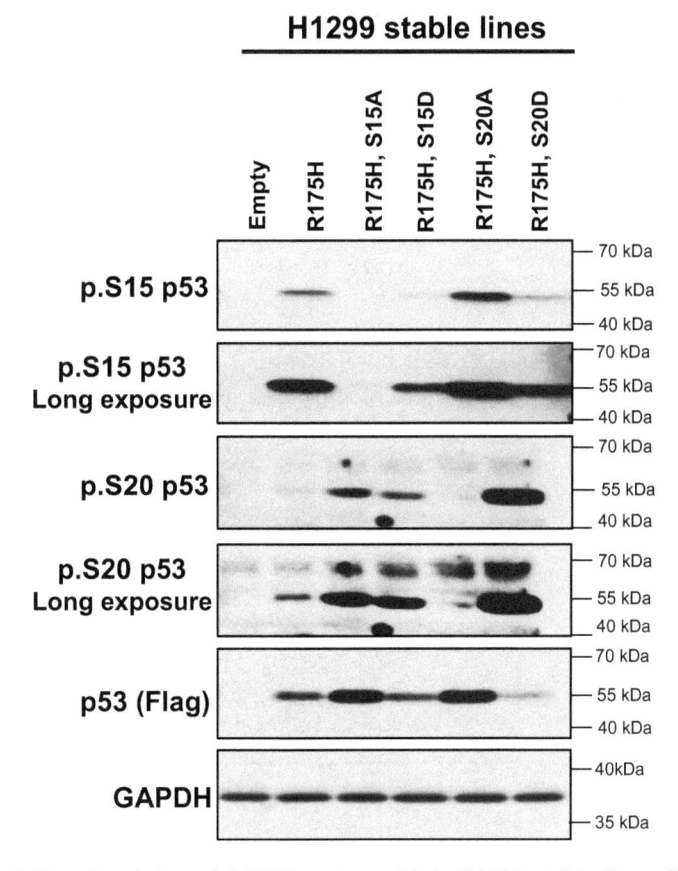

Fig. 4 Phosphorylation of R175H mutant p53 in H1299 stable lines. Serine 15 and serine 20 phosphorylation were evaluated in H1299-based stable lines expressing serine 15 and serine 20 phospho-mimic (i.e., serine to aspartic acid substitution) and phospho-blocking (i.e., serine to alanine substitution) versions of flag-tagged R175H mutant p53. Note that clones express different levels of mutant p53 proteins. Interestingly, when serine 15 and serine 20 were substituted by aspartate (i.e., as phospho-mimic substitutions), R175H mutant p53 could still be recognized by the respective phosphorylation-site specific antibodies, perhaps suggesting that the specificity of the antibodies requires negative charge at these sites

2. The use of milk is not *normally* recommended for the detection of phospho-proteins as milk contains casein, which is a phospho-protein that may be detected by, or interfere with, the phospho-specific antibody resulting in unspecific bands. We have found, however, that using BSA as a blocking agent gives rise to a much higher background, probably due to insufficient blocking of the membrane. We therefore use milk for the blocking process. Also, TBS-T should be used as a diluent, as opposed to PBS-T, for the detection of phosphorylated proteins, for the reason that the high concentration of phosphate in PBS can block the interaction of phospho-specific antibodies with their epitopes.

3. To prepare 5% BSA, weigh 2.5 g and top up to 50 ml with TBS-T using a 50 ml tube. Leave the tube on the roller until dissolved, then filter the solution using filters of 0.45 μm pore size. Filtration will reduce the appearance of tiny dark spots on the blot during development.

4. Antibody dilution buffer and concentrations were optimized to ensure that only phosphorylated p53 is detected and to minimize the detection of unspecific bands. Milk is normally used for high specificity/strongly binding antibodies as it reduces the background. For example, the phospho-serine 15 antibody binds with high affinity and we have found that it works better with milk (same for phosphorylated H2AX). BSA was used for phospho-serine 20 and acetyl-lysine 382 antibodies as they exhibit a relatively weaker affinity. The use of phospho-mimic and phospho-blocking mutant p53 proteins was important for verifying the phospho-specificity of the indicated antibodies (*see* Table 1 for the supplier and catalog number). In Figs. 2, 3, and 4, we used cell lines expressing mutant p53 endogenously (SKBR3, p53-R175H) or, alternatively, ectopically express R175H-mutant p53, or phosphorylation-site mutant derivatives of R175H-mutant p53, to explore specificity of the antibodies (p53-null H1299 cells). We highlight the importance of verifying the phospho-specificity of the antibodies as we have found that some commercially available phospho-p53 antibodies also detect non-modified p53. Mutants were obtained through site-directed mutagenesis (QuickChange Lightning site-directed mutagenesis kit from Agilent Technologies).

5. To stimulate the presence of PTMs on p53, cells can be treated with DNA-damaging agents such as doxorubicin (0.02–1 μM) or etoposide (10–50 μM). Cells can then be harvested in a time frame between 1 and 24 h after treatment. Figures 2 and 3 show DNA damage-induced phosphorylation and acetylation of R175H mutant p53.

6. SDS tends to precipitate at temperatures below 4 °C. Before being sonicated, centrifuged, or loaded on the gel, lysates that have been previously kept on ice should be warmed by leaving the tube for few minutes at room temperature and should be properly mixed afterwards by vortexing.

7. We normally add 0.1 μl of 1 M DTT (stock solution, stored at −20 °C) per μl of protein extract to achieve 0.1 M DTT.

8. We normally use 8% to allow separation of any unspecific bands that may appear, especially when using the anti-phospho serine 20 antibody. 10% gels can also be used.

9. We routinely load 4 μl of PageRuler™ Prestained Protein Ladder (ThermoFisher Scientific) on each gel.

10. After placing the membrane on the gel, ensure that there are no air bubbles between the gel and the membrane. Air bubbles can be removed by rolling them out with a piece of a 5 or 10 ml pipette.

11. When connecting to the power supplier, always ensure that the direction of the current is appropriate.

12. Ponceau staining reversibly visualizes the presence of protein on the membrane and can therefore be used to check the success of the transfer and detect the presence of any gaps that may have arisen if bubbles remained trapped during the transfer. Note if any gap is present around 55 kDa, it can affect the interpretation of the data.

13. Primary antibody solution can be stored at −20 °C in a 15 ml tube and re-used 2–3 times.

14. We use a ChemiDoc™ MP imaging system, which detects the chemiluminescence emerging from the membrane and transforms the signal into a digital image for rapid analysis using the Image Lab 4.1 software. In addition to quantification, this system allows multiple exposure times, from which we are able to select optimal images for presentation. When using the antibodies that recognizes phosphorylated serine 15 or anti-p53 (protein), short exposure times are normally required (from 1 to 200 s). In contrast, detection using phospho-serine 20 and acetyl-lysine 382 antibodies requires longer exposure times (from 200 to 600 s).

Acknowledgments

This work was supported by a studentship from Breast Cancer Now (UK), grant number: 2013NovPhD134.

References

1. Bieging KT, Mello SS et al (2014) Unravelling mechanisms of p53-mediated tumour suppression. Nat Rev Cancer 14:359–370
2. Vousden KH, Lane DP (2007) p53 in health and disease. Nat Rev Mol Cell Biol 8:275–283
3. Appella E, Anderson CW (2001) Post-translational modifications and activation of p53 by genotoxic stresses. Eur J Biochem 268:2764–2772
4. Dai C, Gu W (2010) p53 post-translational modification: deregulated in tumorigenesis. Trends Mol Med 16:528–536
5. Meek DW, Anderson CW (2009) Posttranslational modification of p53: cooperative integrators of function. Cold Spring Harb Perspect Biol 1:a000950
6. Meek DW (2015) Regulation of the p53 response and its relationship to cancer. Biochem J 469:325–346
7. Olivier M, Hollstein M et al (2010) TP53 mutations in human cancers: origins, consequences, and clinical use. Cold Spring Harb Perspect Biol 2:a001008

8. Brosh R, Rotter V (2009) When mutants gain new powers: news from the mutant p53 field. Nat Rev Cancer 9:701–713
9. Freed-Pastor WA, Prives C (2012) Mutant p53: one name, many proteins. Genes Dev 26:1268–1286
10. Suh YA, Post SM et al (2011) Multiple stress signals activate mutant p53 in vivo. Cancer Res 71:7168–7175
11. Nguyen T-A, Menendez D et al (2014) Mutant TP53 posttranslational modifications: challenges and opportunities. Human Mutat 35:738–755

Chapter 3

Global Analyses to Identify Direct Transcriptional Targets of p53

Matthew D. Galbraith, Zdenek Andrysik, Kelly D. Sullivan, and Joaquín M. Espinosa

Abstract

The transcription factor p53 controls a gene expression program with pleiotropic effects on cell biology including cell cycle arrest and apoptosis. Identifying direct p53 target genes within this network and determining how they influence cell fate decisions downstream of p53 activation is a prerequisite for designing therapeutic approaches that target p53 to effectively kill cancer cells. Here we describe a comprehensive multi-omics approach for identifying genes that are direct transcriptional targets of p53. We provide detailed procedures for measuring global RNA polymerase activity, defining p53 binding sites across the genome, and quantifying changes in steady-state mRNA in response to p53 activation.

Key words p53, Nutlin, GRO-seq, ChIP-seq, RNA-seq, Transcription, Transcriptome, Gene Expression, Sequencing

1 Introduction

The transcription factor p53 controls a vast tumor-suppressive gene expression network that orchestrates pleiotropic effects in response to various forms of cellular stress such as DNA damage and oncogene activation [1–5]. The importance of p53 as a tumor suppressor is underscored by the fact that mutations in the TP53 locus occur in about half of human tumors, while in the other half p53 activity is inactivated by other means—such as hyperactivation of its endogenous repressors, MDM2 and MDM4 [6]. In the latter scenario, the potential for reactivation of p53 to serve as a therapeutic approach in cancer treatment has led to the development of MDM2 and MDM4 inhibitors that activate p53 without the genotoxic side effects of traditional chemotherapeutics [7]. However, the outcome of p53 reactivation in most cancer cell types is reversible cell cycle arrest [8], a less desirable therapeutic outcome than cell death. Thus, a better understanding of the influence of direct and indirect p53 target genes on context-dependent cell fate choice

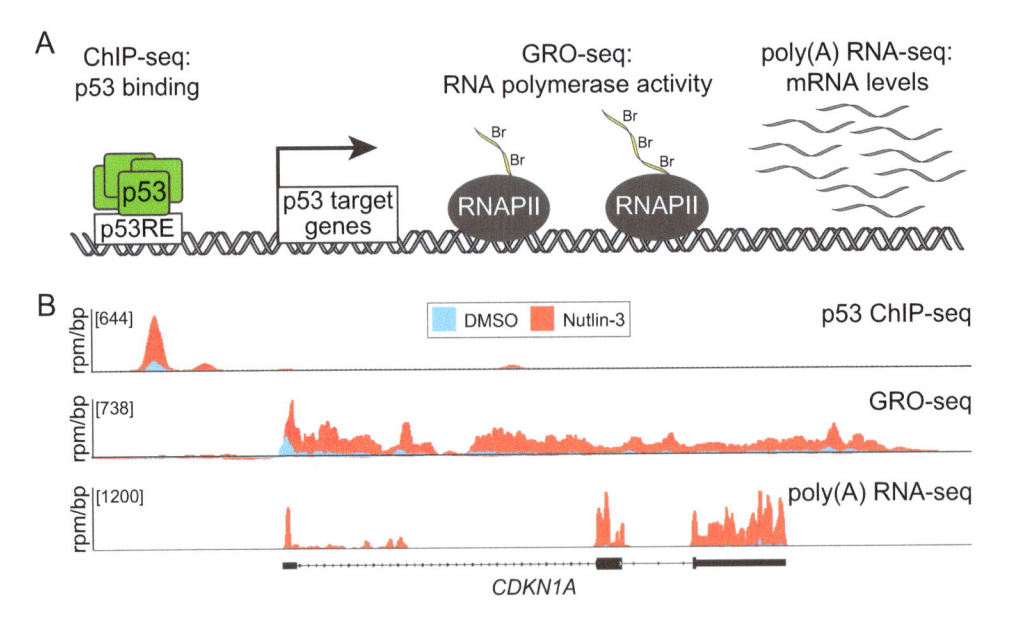

Fig. 1 (**a**) Schematic of multi-omics approach for identifying direct transcriptional targets of p53. Yellow ribbons represent nascent Br-UTP-labeled (Br) RNA produced during the nuclear run-on reaction by RNA polymerase II (RNAPII). Gray ribbons represent steady-state polyadenylated (poly(A)) mRNAs. p53RE, p53 response element, i.e., binding site (**b**) Genome browser views of data for the well-known direct p53 target gene *CDKN1A*, which encodes the cyclin-dependent kinase inhibitor p21

is required. To address this issue, we recently carried out a systematic comparison of the effects of p53 on gene expression at multiple steps of the central dogma in three cancer cell types with differing responses to p53 activation. Using this approach, we identified a set of core direct p53 target genes that are activated regardless of cell context, as well as sets of indirect p53 target genes that are highly divergent across cell types [9]. Here, we describe our procedures for generation of the genomic datasets that allowed us to identify these direct and indirect p53 target genes. Our approach employs global run-on sequencing (GRO-seq) to measure RNA polymerase activity genome-wide, chromatin immunoprecipitation sequencing (ChIP-seq) to define p53 chromatin-binding sites, and RNA sequencing (RNA-seq) to quantify steady-state RNA levels, upon activation of p53 using Nutlin-3 (Fig. 1).

Nutlin-3 is a small molecule inhibitor of the MDM2-p53 interaction [10], which activates p53 both by unmasking the p53 transactivation domains, and by preventing MDM2-dependent p53 ubiquitination and subsequent proteasomal degradation. Activation of p53 using Nutlin-3 is rapid, with at least some activation apparent prior to p53 protein accumulation, and highly-specific, with negligible gene expression changes observed in p53-null cells [10, 11].

The GRO-seq method relies on the high stability and processive nature of transcriptionally engaged RNA polymerases [12]. Nuclei are isolated from cells and washed to remove ribonucleotide triphosphates (rNTPs) and halt RNA synthesis. Transcription is then restarted by adding back rNTPs, substituting Br-UTP for UTP. Following RNA fragmentation, Br-UTP-labeled nascent RNAs are enriched by immunoprecipitation and sequenced, giving a readout of the amount and location of transcriptionally engaged RNA polymerases genome-wide [12]. Thus, at short time points, GRO-seq is ideally suited for measuring the immediate impact of p53 activation on RNA polymerase activity [11].

The ChIP-seq approach relies on sequencing of DNA recovered from fragmented, formaldehyde-cross-linked chromatin by immunoprecipitation using an antibody against the protein of interest [13, 14]. The short-range nature of the protein-DNA cross-linking results in enrichment above background of DNA fragments corresponding to specific binding sites, which can subsequently be identified bioinformatically [15–17]. ChIP-seq is thus able to provide a comprehensive catalog of the genomic binding locations of chromatin-associated proteins and DNA-binding proteins, such as transcription factors, and has previously been used to identify p53 binding sites genome-wide [4, 9, 11].

RNA-seq can be used to monitor a variety of different RNA entities, from unprocessed primary transcripts to small RNAs and mature spliced mRNAs, depending on the sample extraction and library preparation approach used [18]. All methods generally rely on some combination of fragmentation, cDNA synthesis, and adaptor ligation to produce sequencing libraries that can then be analyzed to infer the state of the transcriptome. Here, we describe our protocol for the measurement of steady-state mRNA levels after activation of p53 with Nutlin-3.

Comparison of these three datasets allows the comprehensive identification of genes that are directly and indirectly regulated by p53. For example, we define direct target genes as those that: (1) display increases in GRO-seq signals at short time points upon p53 activation and (2) show increased RNA-seq signals and an associated p53 binding site after p53 activation. In contrast, we define indirect target genes as those that display an increase in mRNA levels but lack an associated p53 binding site [9].

2 Materials

Prepare all solutions using ultrapure, nuclease-free water. Take precautions to ensure nuclease-free conditions throughout all procedures. All solutions/reagents should be stored at room temperature (RT) unless otherwise specified.

2.1 Cell Culture and Activation of p53 by Nutlin Treatment

1. Cell culture medium and supplements suitable for growth of chosen cells.
2. Tissue culture plates suitable for growth of chosen cells, 10 cm ø (~58 cm^2) and 15 cm ø (~143 cm^2) (Greiner, 664160 and 639160, or similar).
3. Sterile phosphate-buffered saline (PBS): 137 mM NaCl, 2.7 mM KCl, 10 mM Na$_2$HPO$_4$, 1.8 mM KH$_2$PO$_4$. Adjust pH to 7.4 with HCl and autoclave prior to use.
4. 0.25% Trypsin-EDTA solution (ThermoFisher Scientific, 25200056, or similar).
5. Hemocytometer or similar for counting cells.
6. Dimethyl sulfoxide (DMSO), tissue culture grade.
7. Nutlin-3a (Cayman Chemical, 18585, or similar): dissolve in DMSO to make a 1000× stock concentration at 10 mM for use at 10 µM final concentration (*see* **Note 1**). Store in small aliquots at −20 °C.

2.2 Isolation of Nuclei for GRO-Seq

1. ~1 × 10^8 sub-confluent cells per condition/treatment, cultured in 15 cm ø tissue culture plates (~1 × 10^7 per plate).
2. Diethylpyrocarbonate (DEPC, Sigma Aldrich, D5758).
3. DEPC-treated H$_2$O: treat with 0.1% (v/v) DEPC for 2 h at 37 °C and autoclave thereafter. Alternatively, buy nuclease-free water.
4. Phosphate-buffered saline (PBS).
5. SUPERase-In RNase inhibitor (ThermoFisher Scientific, AM2694).
6. Lysis buffer: 10 mM Tris–HCl pH 7.4, 2 mM MgCl$_2$, 3 mM CaCl$_2$, 0.5% (w/v) IGEPAL CA-630 (NP-40 substitute), 10% (w/v) glycerol, 1 mM DTT, protease inhibitor cocktail (cOmplete Mini, Roche, 11836170001), 4 U/mL SUPERase-In. Make up in DEPC-treated H$_2$O. Store at 4 °C. Add DTT, protease inhibitor cocktail, and SUPERase-In immediately before use.
7. 50 mL conical tubes.
8. Large orifice P200 tips (USA Scientific, 1011-8410).
9. Prelubricated microcentrifuge tubes, 1.7 mL (Costar, 3207) (*see* **Note 2**).
10. DNA LoBind microcentrifuge tubes, 0.5, 1.5 mL, 2 mL (Eppendorf, 022431005, 022431021, 022431048), or similar (*see* **Note 3**).
11. Trypan blue dye: 0.4% (w/v) trypan blue, 0.85% (w/v) NaCl (Lonza, 17-942E), store at RT. Dilute 1:2 with PBS and 0.2 µm filter before use.

12. Hemocytometer for counting nuclei.
13. Freezing buffer: 50 mM Tris–HCl pH 8.3, 5 mM MgCl$_2$, 40% (w/v) glycerol, 0.1 mM EDTA, 4 U/mL SUPERase-In. Make up in DEPC-treated H$_2$O. Store at 4 °C. Add SUPERase-In immediately before use.

2.3 Nuclear Run-On and Nascent RNA Enrichment for GRO-Seq

1. 5×10^6 nuclei per reaction, prepared as in Subheading 3.2.1.
2. 2 M Tris–HCl pH 8.0.
3. 1 M MgCl$_2$.
4. 0.5 M Dithiothreitol (DTT). Always prepare fresh in nuclease-free H$_2$O, do not store.
5. 3 M KCl.
6. 100 mM rATP, rGTP, rCTP (Promega, E6018, E6038, E6048). Dilute to 10 mM with DEPC-treated H$_2$O and store small aliquots at −20 °C.
7. 5-Bromouridine 5′-triphosphate sodium salt (Br-UTP) (Sigma Aldrich, B7166). Make up 10 mM solution in DEPC-treated H$_2$O. Store small aliquots at −20 °C.
8. DEPC-treated H$_2$O.
9. DNA LoBind microcentrifuge tubes, 0.5, 1.5 mL, 2 mL (Eppendorf, 022431005, 022431021, 022431048), or similar (*see* **Note 3**).
10. 30 °C water bath.
11. SUPERase-In RNase inhibitor (ThermoFisher Scientific, AM2694).
12. *N*-Lauroylsarcosine sodium salt (sarkosyl, Sigma Aldrich, L5777). Make up 2% (w/v) solution in DEPC-treated H$_2$O. Store at 4 °C.
13. Large orifice P200 tips (USA Scientific, 1011-8410).
14. TRIzol reagent or similar (Invitrogen/ThermoFisher Scientific, 15,596,026), store at 4 °C.
15. Acid phenol:chloroform:isoamyl alcohol 25:24:1 pH 4.5 (Invitrogen/ThermoFisher Scientific, AM9720), store at 4 °C.
16. Chloroform.
17. GlycoBlue (Invitrogen/ThermoFisher Scientific, AM9515), store at −20 °C.
18. DEPC-treated H$_2$O.
19. 1 M NaOH, store 1 mL aliquots at −20 °C.
20. 1 M Tris–HCl pH 6.8.

21. RNase-free Micro Bio-Spin P-30 columns, Tris Buffer (7326250, BioRad), store at 4 °C.
22. Antarctic Phosphatase (M0289S, New England Biolabs), store at −20 °C.
23. RQ1 RNase-free DNase (Promega, M610A), store at −20 °C.
24. Anti-BrdU agarose beads (sc-32323 AC, Santa Cruz Biotechnology), store at 4 °C. Alternatively, anti-BrdU antibodies from other vendors can be used in combination of beads of choice. However, it is absolutely neccesary to test the immunoprecipitation efficiency of the substitute reagents by qRT-PCR before using them in the GRO-seq protocol.
25. Rotating platform or similar for mixing tubes.
26. 20× SSPE buffer: 3 M NaCl, 0.2 M NaH_2PO_4, 20 mM EDTA, adjusted to pH 7.4 with 10 M NaOH, 0.2–0.45 μm filtered.
27. Binding buffer: 0.25× SSPE, 1 mM EDTA, 0.05% (v/v) Tween-20, 37.5 mM NaCl, 4 U/mL SUPERase-In. Prepare fresh before use, do not store. Add SUPERase-In immediately before use.
28. Blocking buffer: 0.25× SSPE, 1 mM EDTA, 0.05% (v/v) Tween-20, 37.5 mM NaCl, 0.1% (w/v) Polyvinylpyrrolidone (Sigma Aldrich, PVP40), 1 μg/mL BSA (New England Biolabs, B9000S), 4 U/mL SUPERase-In. Prepare fresh before use, do not store. Add SUPERase-In immediately before use.
29. Low Salt Wash buffer: 0.2× SSPE, 1 mM EDTA, 0.05% (v/v) Tween-20, 4 U/mL SUPERase-In. Store <1 week at 4 °C. Add SUPERase-In immediately before use.
30. High Salt Wash buffer: 0.25× SSPE, 1 mM EDTA, 0.05% (v/v) Tween-20, 137.5 mM NaCl, 4 U/mL SUPERase-In. Store <1 week at 4 °C. Add SUPERase-In immediately before use.
31. TET buffer: 10 mM Tris pH 8, 1 mM EDTA, 0.05% (v/v) Tween-20, 4 U/mL SUPERase-In. Store <1 week at 4 °C. Add SUPERase-In immediately before use.
32. GRO-seq Elution buffer: 20 mM DTT, 1 mM EDTA, 150 mM NaCl, 50 mM Tris–HCl pH 7.5, 0.1% (w/v) SDS. Prepare fresh before use, do not store.
33. Phenol:chloroform:isoamyl alcohol 25:24:1 pH 8 (Sigma Aldrich, 77617), store at 4 °C.
34. 5 M NaCl.
35. Ethanol, store and use at −20 °C.

2.4 Sequencing Library Preparation for GRO-Seq

1. Water bath(s) or heating block(s) at 30 °C, 37 °C, and 42 °C.
2. DNA LoBind microcentrifuge tubes, 0.5, 1.5 mL, 2 mL (Eppendorf, 022431005, 022431021, 022431048), or similar (see **Note 3**).
3. T4 polynucleotide kinase (New England Biolabs, M0201S), store at −20 °C.
4. Poly(A) polymerase (New England Biolabs, M0276S), store at −20 °C.
5. 0.5 M EDTA.
6. 5 M NaCl.
7. GRO-seq RT primer, reconstitute at 12.5 µM in nuclease-free H_2O, store at −20 °C see Table 1.
8. dNTPs 100 mM (ThermoFisher Scientific, R0181). Prepare a combined solution at 10 mM each in nuclease-free H_2O and store small aliquots at −20 °C.
9. SUPERase-In RNase inhibitor (ThermoFisher Scientific, AM2694).
10. 1 M DTT. Always prepare fresh in nuclease-free H_2O, do not store.
11. 1 M $MgCl_2$.
12. Superscript III reverse transcriptase (Invitrogen/ThermoFisher Scientific, 18080051), store at −20 °C.
13. Exonuclease I (ThermoFisher Scientific, EN0581), store at −20 °C.
14. 1 M NaOH.
15. 1 M HCl.
16. Phenol:chloroform:isoamyl alcohol 25:24:1 pH 8 (Sigma Aldrich, 77617), store at 4 °C.
17. Chloroform.
18. GlycoBlue (Invitrogen/ThermoFisher, AM9515), store at −20 °C.
19. Ethanol, store and use at −20 °C.
20. Nuclease-free H_2O.
21. BluePippin electrophoretic fractionator (Sage Science).
22. 2% agarose gel cassette kit (Sage Science, BDF2010).
23. CircLigase ssDNA ligase (Epicenter/Lucigen, CL4111K), store at −20 °C.
24. APE 1 (New England Biolabs, M0282S), store at −20 °C.
25. Phusion high-fidelity DNA polymerase (ThermoFisher Scientific, F530), store at −20 °C.
26. RNA PCR Index primers, reconstitute at 25 µM in nuclease-free H_2O, store at −20 °C see Table 1.

Table 1
GRO-seq library primer sequences

Procedure	Primer/Adapter	Sequence (5' - 3')
Reverse transcription	GRO-seq RT primer	**p**GATCGTCGGACTGTAGAACTCT/**idSp**/CCTTGGCACCCGAGAATTCCATTTTTTTTTTTTTTTTTTT**VN**
Library amplification[a]	RP1	AATGATACGGCGACCACCGAGATCTACACGTTCAGAGTTCTACAGTCCGA
	RPI1	CAAGCAGAAGACGGCATACGAGAT**CGTGAT**GTGACTGGAGTTCCTTGGCACCCGAGAATTCCA
	RPI2	CAAGCAGAAGACGGCATACGAGAT**ACATCG**GTGACTGGAGTTCCTTGGCACCCGAGAATTCCA
	RPI3	CAAGCAGAAGACGGCATACGAGAT**GCCTAA**GTGACTGGAGTTCCTTGGCACCCGAGAATTCCA
	RPI4	CAAGCAGAAGACGGCATACGAGAT**TGGTCA**GTGACTGGAGTTCCTTGGCACCCGAGAATTCCA
	RPI5	CAAGCAGAAGACGGCATACGAGAT**CACTGT**GTGACTGGAGTTCCTTGGCACCCGAGAATTCCA
	RPI6	CAAGCAGAAGACGGCATACGAGAT**ATTGGC**GTGACTGGAGTTCCTTGGCACCCGAGAATTCCA
	RPI7-48	CAAGCAGAAGACGGCATACGAGAT**NNNNNN** GTGACTGGAGTTCCTTGGCACCCGAGAATTCCA

[a]These primers are based on the Illumina TruSeq small RNA PCR Index primers (RPI). NNNN and NNNN = sequence complementarity between RT primer and RPI primers, NNNN = six base index/barcode. Abbreviations: p = 5' phosphorylation, /idSp/ = abasic furan internal dSpacer, V = degenerate A/C/G nucleotide, N = A/G/C/T degenerate nucleotide.

27. 5× Orange G buffer: 50% (v/v) glycerol, 0.5% (w/v) Orange G, 5 mM EDTA.
28. 1 kb Plus DNA ladder (ThermoFisher Scientific, 10787018), or similar. Store at −20 °C.
29. Ethidium bromide.

2.5 Quantification and Pooling of GRO-Seq Libraries for Multiplex Sequencing

1. Qubit fluorometer (ThermoFisher Scientific).
2. Qubit dsDNA HS assay kit (ThermoFisher Scientific).
3. 2100 Bioanalyzer (Agilent).
4. Bioanalyzer High-Sensitivity DNA analysis kit (Agilent).
5. Nuclease-free H_2O.

2.6 Cross-Linking and Whole-Cell Lysate Preparation for ChIP-Seq

1. ~5 × 10^7 sub-confluent cells per condition/treatment, cultured in 15 cm ø tissue culture plates (~1 × 10^7 per plate).
2. PBS (*see* **Note 5**).
3. 1% (v/v) formaldehyde in PBS. Prepare with RT PBS immediately before use (*see* **Note 6**).
4. 2.5 M glycine, 0.2–0.45 μm filtered (*see* **Note 7**).
5. RIPA buffer: 150 mM NaCl, 50 mM Tris–HCl pH 8.0, 5 mM EDTA, 1% (v/v) IGEPAL CA-630 (NP-40 substitute), 0.5% (w/v) sodium deoxycholate, 0.1% (w/v) SDS. 0.2–0.45 μm filtered. Store at 4 °C. For complete RIPA buffer, add protease, phosphatase, and/or histone deacetylase inhibitors immediately before use (*see* **Note 8**).
6. Large cell scraper(s).
7. 15 mL conical tubes.
8. Probe or bath sonicator, or similar for chromatin fragmentation.
9. 5× Orange G buffer: 50% (v/v) glycerol, 0.5% (w/v) Orange G, 5 mM EDTA.
10. 1 kb Plus DNA ladder (ThermoFisher Scientific, 10787018), or similar. Store at −20 °C.
11. Ethidium bromide.
12. 30 mL centrifuge tubes rated for 12,000 × *g* and suitable refrigerated centrifuge.
13. BCA protein assay kit (Pierce/ThermoFisher Scientific, 23227), includes bovine serum albumin (BSA) standard.
14. ~1.7 mL microcentrifuge tubes.
15. Liquid nitrogen for snap freezing lysates.

2.7 Immunoprecipitation and DNA Purification for ChIP-Seq

1. Sheep anti-mouse Dynabeads M-280 (Invitrogen/ThermoFisher Scientific, 11201D).
2. Magnetic separator rack, e.g., DynaMag (ThermoFisher Scientific, 12320D/12321D) or similar (*see* **Note 9**).
3. Mouse monoclonal anti-p53 antibody, clone DO-1 (MilliporeSigma, OP43).

4. DNA LoBind microcentrifuge tubes, 0.5, 1.5 mL, 2 mL (Eppendorf, 022431005, 022431021, 022431048), or similar (*see* **Note 3**).

5. RIPA buffer: 150 mM NaCl, 50 mM Tris–HCl pH 8.0, 5 mM EDTA, 1% (v/v) IGEPAL CA-630 (NP-40 substitute), 0.5% (w/v) sodium deoxycholate, 0.1% (w/v) SDS. 0.2–0.45 μm filtered. Store at 4 °C. For complete RIPA buffer, add protease, phosphatase, and/or histone deacetylase inhibitors immediately before use (*see* **Note 8**).

6. IP wash buffer: 500 mM LiCl, 100 mM Tris–HCl pH 8.5, 1% (v/v) IGEPAL CA-630 (NP-40 substitute), 1% (w/v) sodium deoxycholate. 0.2–0.45 μm filtered. Store at 4 °C.

7. TE buffer: 10 mM Tris–HCl pH 8, 1 mM EDTA.

8. 1.5× ChIP Elution buffer: 70 mM Tris–HCl pH 8, 1 mM EDTA, 1.5% (w/v) SDS.

9. 5 M NaCl.

10. 20 mg/mL Proteinase K (Sigma Aldrich, P6556).

11. Phenol:chloroform:isoamyl alcohol 25:24:1 pH 8 (Sigma Aldrich, 77617), store at 4 °C.

12. Chloroform.

13. GlycoBlue (Invitrogen/ThermoFisher Scientific, AM9515), store at −20 °C.

14. 3 M sodium acetate pH 4.8.

15. Ethanol.

2.8 Size Selection and Sequencing Library Preparation for ChIP-Seq

1. Qubit fluorometer (ThermoFisher Scientific).
2. Qubit dsDNA HS assay kit (ThermoFisher Scientific, Q32851).
3. BluePippin electrophoretic fractionator (Sage Science).
4. 2% agarose gel cassette kit (Sage Science, BDF2010).
5. TruSeq ChIP Library preparation kit (Illumina), or similar.

2.9 Quantification and Pooling of ChIP-Seq Libraries for Multiplex Sequencing

1. Qubit fluorometer (ThermoFisher Scientific).
2. Qubit dsDNA HS assay kit (ThermoFisher Scientific, Q32851).
3. 2100 Bioanalyzer (Agilent).
4. Bioanalyzer High Sensitivity DNA analysis kit (Agilent, 5067-4626).
5. Nuclease-free water.

2.10 Isolation of Total Cellular RNA for RNA-Seq

1. ~5×10^6 sub-confluent cells per condition/treatment, cultured in 10 cm ø tissue culture plates (~5×10^6 per plate).
2. PBS, store at 4 °C.
3. TRIzol reagent or similar (Invitrogen/ThermoFisher Scientific, 15596026), store at 4 °C.
4. DNA LoBind microcentrifuge tubes, 0.5, 1.5 mL, 2 mL (Eppendorf, 022431005, 022431021, 022431048), or similar (*see* **Note 3**).
5. Chloroform.
6. Isopropanol.
7. Ethanol.
8. Nuclease-free H_2O.
9. Synergy 2 Take 3 module (Biotek), NanoDrop (ThermoFisher), or similar.
10. Qubit fluorometer (ThermoFisher Scientific).
11. Qubit RNA BR or HS assay kit (ThermoFisher Scientific, Q10210 or Q32852).
12. 2100 Bioanalyzer (Agilent).
13. RNA 6000 Pico assay kit (Agilent, 5067-1513).

2.11 Poly(A) Enrichment and Sequencing Library Preparation for RNA-Seq

1. Dynabeads mRNA Direct micro kit (Ambion/ThermoFisher Scientific, 61021) or similar, if using sequencing library preparation kit that does not include poly(A) enrichment.
2. TruSeq RNA Library Preparation Kit v2 (Illumina, RS-122-2001 or RS-122-2002), or similar for Illumina sequencing. Ion Total RNA-seq kit v2 for Ion Torrent sequencing (ThermoFisher Scientific, 4475936 or 4479789). We do not recommend using library preparation methods that are not strand-specific.
3. Qubit fluorometer (ThermoFisher Scientific).
4. Qubit dsDNA HS assay kit (ThermoFisher Scientific, Q32851).
5. 2100 Bioanalyzer (Agilent).
6. Bioanalyzer High Sensitivity DNA analysis kit (Agilent, 5067-4626).
7. Nuclease-free water.

2.12 Computational Tools

The bioinformatic analyses described here should be performed in a Unix-based environment such as Linux or Mac OS X, and will require some knowledge of command line tools and languages such as BASH, Python, and R. Access to a high-performance computing system will greatly reduce the time required for most analysis steps. In general, we recommend using the latest version of each tool.

1. FastQC https://www.bioinformatics.babraham.ac.uk/projects/fastqc/
2. FastQ Screen https://www.bioinformatics.babraham.ac.uk/projects/fastq_screen/
3. Ea-utils https://expressionanalysis.github.io/ea-utils/
4. FASTX-Toolkit http://hannonlab.cshl.edu/fastx_toolkit/
5. Bowtie 2 http://bowtie-bio.sourceforge.net/bowtie2/index.shtml [19].
6. Human hg19/GRCh37 index for bowtie (from iGenomes UCSC hg19 bundle) https://support.illumina.com/sequencing/sequencing_software/igenome.html.
7. Picard tools http://broadinstitute.github.io/picard/
8. SAMtools http://www.htslib.org/ [20].
9. Pybedtools https://daler.github.io/pybedtools/ [21, 22].
10. HTSeq https://pypi.python.org/pypi/HTSeq [23].
11. Human UCSC hg19 GTF annotation file (from iGenomes UCSC hg19 bundle) https://support.illumina.com/sequencing/sequencing_software/igenome.html.
12. R https://www.r-project.org/
13. Bioconductor http://www.bioconductor.org/ [24].
14. DESeq 2 http://bioconductor.org/packages/release/bioc/html/DESeq2.html [25].
15. HOMER http://homer.ucsd.edu/homer/motif/ [26].
16. deepTools https://deeptools.readthedocs.io/en/latest/ [27].
17. RSeQC http://rseqc.sourceforge.net/ [28].
18. hg19.HouseKeepingGenes.bed file obtained from https://sourceforge.net/projects/rseqc/files/BED/Human_Homo_sapiens/hg19.HouseKeepingGenes.bed/download
19. Python https://www.python.org/

3 Methods

3.1 Cell Culture and Non-genotoxic Activation of p53

1. Ensure that cells are healthy and have not been excessively passaged.
2. Remove media and wash with an equal volume of RT PBS.
3. Remove PBS by aspiration and add a sufficient volume of 0.25% trypsin solution to cover and detach cells. Incubate for ~5 min at 37 °C (*see* **Note 10**).
4. Quench trypsin with a suitable volume of FBS-containing culture medium and thoroughly resuspend cells, ensuring a single-cell suspension.

5. Determine cell concentration using a hemocytometer or similar.
6. Dilute cells and dispense an appropriate number of cells into 10 cm ø and/or 15 cm ø tissue culture plates to give ~80% confluency at the end of treatment (*see* **Note 11**).
7. Incubate cells under standard growth conditions for 16–24 h to allow recovery.
8. To treat cells, add 1/1000th volume of 10 mM Nutlin-3a (for a final concentration of 10 µM) or the equivalent volume of DMSO (as vehicle control) directly to the media in each plate. Gently swirl to mix. Incubate plates for the desired treatment time (*see* **Note 12**).
9. Proceed to harvest cells according to Subheadings 3.2.1, 3.3.1, or 3.4.1.
10. In all cases, at least two biological replicates should be performed.

3.2 Profiling the Transcriptional Response to p53 Activation by Global Run-on Sequencing (GRO-Seq)

3.2.1 Isolation of Nuclei for GRO-Seq

1. Treat cells in 15 cm ø plates with DMSO or Nutlin-3a, as described in Subheading 3.1, to obtain ~1×10^8 cells per condition/treatment at harvest (*see* **Note 11**). Include extra plate(s) to collect parallel protein and RNA samples for quality control and validation by western blotting and qRT-PCR analysis.
2. Discard growth media (*see* **Note 13**).
3. Keep plates on ice through **step 5**.
4. Gently wash cells on plates three times with 20 mL ice-cold PBS, removing all PBS after last wash (*see* **Note 14**).
5. Add 1 mL ice-cold lysis buffer per 15 cm ø plate. Attempt to cover all cells with lysis buffer. Harvest cells by scraping, and pool all cells of the same treatment/condition in a 50 mL conical tube on ice. Rinse plates with another 1 mL ice-cold lysis buffer to recover remaining cells, and add to the 50 mL conical tube.
6. Pellet cells by centrifugation at $1000 \times g$ for 15 min at 4 °C.
7. Resuspend in 1.5 mL ice-cold lysis buffer by pipetting 20–30 times with a P1000 tip to lyse cell membranes and release intact nuclei.
8. Add a further 8.5 mL ice-cold lysis buffer and pellet nuclei by centrifugation at $1000 \times g$ for 15 min at 4 °C (*see* **Note 15**).
9. Wash nuclei by resuspending in 1 mL ice-cold lysis buffer using a large orifice P200 tip and transfer to a prelubricated 1.7 mL tube, and pellet by centrifugation at $1000 \times g$ for 5 min at 4 °C.
10. Resuspend nuclei in 500 µL ice-cold freezing buffer using a large orifice P200 tip (*see* **Note 16**).

11. Pellet nuclei by centrifugation at 2000 × *g* for 2 min at 4 °C.
12. Resuspend nuclei in 110 μL ice-cold freezing buffer using a large orifice P200 tip (*see* **Note 16**).
13. Determine concentration and yield of nuclei by making a 1:50 dilution by adding 2 μL of the nuclei suspension to 98 μL of 1× trypan blue in PBS, and counting on a hemocytometer (*see* **Note 17**).
14. Label and pre-cool an appropriate number of prelubricated 1.7 mL tubes on ice.
15. Dilute nuclei with ice-cold freezing buffer as necessary to make 100 μL aliquots of 5×10^6 nuclei in prelubricated 1.7 mL tubes.
16. Snap freeze in liquid nitrogen and store at −80 °C.

3.2.2 Nuclear Run-on and RNA Extraction

1. Make up 2× master mix according to Table 2, without SUPERase-In or sarkosyl (*see* **Note 18**).
2. In one tube per reaction, add reagents in the following order and preheat in 30 °C water bath: 49 μL 2× master mix, 1 μL SUPERase-In, 50 μL 2% sarkosyl.
3. Initiate run-on reactions by adding a 100 μL aliquot of nuclei, mixing (*see* **Note 19**), and incubate for 5 min at 30 °C (*see* **Note 20**). Flick tubes to mix at 2 and 4 min.
4. Stop reactions by adding 1 mL TRIzol reagent and mixing thoroughly (*see* **Note 21**). Divide into two separate tubes (600 μL each). Add an additional 500 μL TRIzol reagent and mix thoroughly.

Table 2
Composition of 2× Reaction mix for nuclear run-on

Reagent	Final reaction concentration	Volume in 2× Master mix per reaction
2 M Tris–HCl pH 8.0	10 mM	0.5 μL
1 M MgCl$_2$	5 mM	0.5 μL
0.5 M DTT	1 mM	0.2 μL
3 M KCl	300 mM	10 μL
10 mM rATP	0.5 mM	5 μL
10 mM rGTP	0.5 mM	5 μL
10 mM rCTP	0.5 mM	5 μL
10 mM Bromo-UTP	0.5 mM	5 μL
DEPC-treated H$_2$O	–	17.8 μL
Total volume	–	49 μL

5. Add 200 μL chloroform per mL of TRIzol reagent to each tube, shake for 15 s, and let stand for 10 min at RT. Centrifuge at 12,000 × *g* for 15 min at 4 °C. Transfer aqueous layers to a new tube.

6. Add ~22.5 μL 5 M NaCl (~300 mM Cl⁻ ions final, including KCl in master mix).

7. Add an equal volume of RT acid phenol:chloroform, vortex for 10 s, and centrifuge at 12,000 × *g* for 10 min at room temperature. Transfer each aqueous layer to a new tube.

8. Add an equal volume of chloroform, vortex for 10 s, and centrifuge at 12,000 × *g* for 10 min at room temperature. Transfer each aqueous layer to a new tube.

9. Add 1/100th volume of GlycoBlue to each sample and precipitate RNA by adding three volumes of ice-cold ethanol.

10. Incubate for 20 min at −20 °C (*see* **Note 22**).

11. Centrifuge precipitated RNA at ≥15,000 × *g* for 20 min at 4 °C, and discard supernatant.

12. Wash RNA pellet with 1 mL 75% (v/v) ethanol, centrifuge at ≥15,000 × *g* for 5 min, and discard supernatant.

13. Air-dry the RNA pellet for 2 min at RT (*see* **Note 23**), resuspend in 20 μL DEPC-treated H$_2$O, and keep on ice. If multiple reactions per treatment/condition were performed, combine all pellets in 20 μL.

3.2.3 Nascent RNA Enrichment and Sequencing Library Preparation for GRO-Seq

1. *RNA fragmentation*: Fragment the nuclear RNA (20 μL) using base hydrolysis by adding 5 μL of 1 M NaOH and incubating on ice for up to 60 min (*see* **Note 24**).

2. Stop/neutralize the hydrolysis reaction by adding ~25 μL 1 M Tris–HCl pH 6.8 (*see* **Note 25**).

3. Desalt the fragmented RNA by passing through an RNase-free P-30 spin column according to the manufacturer's instructions.

4. *DNase treatment*: Measure volume (~60 μL), and add 10× RQ1 buffer for a final 1× concentration (7 μL), followed by 3 μL RQ1 DNase (70 μL total volume). Incubate for 10 min at 37 °C.

5. Desalt a second time as for **step 3**.

6. *Dephosphorylation*: Add 8.5 μL 10× Antarctic Phosphatase buffer, 1 μL SUPERase-In, and 5 μL Antarctic Phosphatase (~85 μL total volume). Incubate for 1 h at 37 °C.

7. Desalt a third time as for **step 3**.

8. Add 1 μL 0.5 M EDTA (5 mM final) and DEPC-treated H$_2$O to bring volume to 100 μL, and keep on ice.

9. *Br-UTP enrichment #1*: For each sample, equilibrate 60 μL anti-BrdU agarose beads by washing in 500 μL binding buffer on a rotator for 5 min at RT (*see* **Note 26**). Centrifuge beads at 1000 × g for 2 min at RT and discard supernatant, taking care not to disturb beads, and repeat wash once.

10. Block beads in 500 μL blocking buffer on a rotator for 1–2 h at RT. Centrifuge beads at 1000 × g for 2 min at RT and discard supernatant. This step can be carried out in parallel with **steps 1–8**.

11. Wash blocked beads twice in 500 μL binding buffer on a rotator for 5 min at RT, as for **step 9**.

12. Resuspend beads in 400 μL binding buffer.

13. Denature the fragmented RNA from **step 8** by heating for 5 min at 65 °C, then return to ice for 2 min.

14. Add the fragmented RNA to the blocked beads from **step 12**, and incubate on a rotator for 1 h at RT.

15. Pellet beads by centrifugation at 1000 × g for 2 min, and discard supernatant.

16. Wash RNA-bound beads once with 500 μL binding buffer on a rotator for 5 min at RT, centrifuge, and discard supernatant.

17. Wash RNA-bound beads once with 500 μL Low Salt Wash buffer on a rotator for 5 min at RT, centrifuge, and discard supernatant.

18. Briefly wash RNA-bound beads once with 500 μL High Salt Wash buffer (resuspend only), centrifuge, and discard supernatant.

19. Wash RNA-bound beads twice with 500 μL TET buffer on a rotator for 5 min at RT, centrifuge, and discard supernatant.

20. Elute RNA by resuspending in 125 μL GRO-seq elution buffer and incubating for 10 min in a water bath at 42 °C. Mix every 2–3 min. Pellet beads by centrifugation at 1000 × g for 2 min, and transfer supernatant to a new 1.5 mL LoBind tube.

21. Repeat **step 19** three more times, pooling the eluates in one tube.

22. Extract the pooled eluted RNA (~500 μL) with one volume of acid phenol:chloroform, centrifuge, and transfer aqueous phase (top) to a new 1.5 mL LoBind tube.

23. Extract with one volume of chloroform, centrifuge, and transfer aqueous phase (top) to a new 2 mL LoBind tube.

24. Add 1/100th volume GlycoBlue and 15 μL 5 M NaCl (~300 mM final, accounting for salt in GRO-seq elution buffer).

25. Precipitate recovered RNA by adding three volumes of cold ethanol and incubating at −20 °C for 20 min (*see* **Note 22**).

26. Centrifuge precipitated RNA at ≥15,000 × *g* for 20 min at 4 °C, and discard supernatant.

27. Wash RNA pellet with 1 mL 75% (v/v) ethanol, centrifuge at ≥15,000 × *g* for 5 min, and discard supernatant.

28. Air-dry the RNA pellet for 2 min at RT, dissolve in 45 μL DEPC-treated H_2O, and keep on ice.

29. *PNK treatment*: 5.2 μL T4 PNK buffer, 1 μL SUPERase-In, and 1 μL T4 PNK. Incubate 1 h at 37 °C.

30. Add 225 μL DEPC-treated H_2O, 5 μL 0.5 M EDTA, and 18 μL 5 M NaCl.

31. Extract the phosphorylated RNA (~500 μL) with one volume of acid phenol:chloroform, centrifuge, and transfer aqueous phase to a new 1.5 mL LoBind tube.

32. Extract with one volume of chloroform, centrifuge, and transfer aqueous phase to a new 1.5 mL LoBind tube.

33. Add 1/100th volume GlycoBlue and precipitate RNA by adding three volumes cold 100% ethanol and incubating at −20 °C for 20 min (*see* **Note 22**).

34. Centrifuge precipitated RNA at ≥15,000 × *g* for 20 min at 4 °C, and discard supernatant.

35. Wash the Br-UTP-enriched RNA pellet with 1 mL 75% (v/v) ethanol, centrifuge at ≥15,000 × *g* for 5 min, and discard supernatant.

36. Air-dry the RNA pellet for 2 min at RT, resuspend in 5 μL DEPC-treated H_2O, and keep on ice.

37. *Poly(A) tailing*: Add 0.8 μL poly(A) polymerase buffer, 0.8 μL 10 mM ATP, 0.5 SUPERase-In, and 0.75 μL poly(A) polymerase. Incubate 10 min at 37 °C (*see* **Note 27**).

38. Stop reaction by adding 12 μL 0.5 M EDTA, and add 12 μL 5 M NaCl and 184 μL DEPC-treated H_2O.

39. Extract the phosphorylated RNA (~216 μL) with one volume of acid phenol:chloroform, centrifuge, and transfer aqueous phase to a new 1.5 mL LoBind tube.

40. Extract with one volume of chloroform, centrifuge, and transfer aqueous phase to a new 1.5 mL LoBind tube.

41. Add 1/100th volume GlycoBlue and precipitate RNA by adding three volumes cold 100% ethanol and incubating at −20 °C for 20 min (*see* **Note 22**).

42. Dissolve recovered RNA in 50 μL DEPC-treated H_2O and add 1 μL 0.5 M EDTA.

43. *Br-UTP enrichment #2*: For each sample, equilibrate 60 μL anti-BrdU agarose beads by washing in 500 μL binding buffer on a rotator for 5 min at RT (*see* **Note 26**). Centrifuge beads at $1000 \times g$ for 2 min at RT and discard supernatant, taking care not to disturb beads, and repeat wash once. This step can be carried out in parallel with **step 41**.

44. Block beads in 500 μL blocking buffer on a rotator for 1–2 h at RT. Centrifuge beads at $1000 \times g$ for 2 min at RT and discard supernatant.

45. Wash blocked beads twice in 500 μL binding buffer on a rotator for 5 min at RT, as for **step 43**.

46. Resuspend beads in 450 μL binding buffer.

47. Denature the recovered RNA from **step 42** by heating for 5 min at 65 °C, then return to ice for 2 min.

48. Add the denatured RNA to the blocked beads from **step 46**, and incubate on a rotator for 1 h at RT.

49. Pellet beads by centrifugation at $1000 \times g$ for 2 min, and discard supernatant.

50. Wash RNA-bound beads once with 500 μL binding buffer on a rotator for 5 min at RT, centrifuge, and discard supernatant.

51. Wash RNA-bound beads once with 500 μL Low Salt Wash buffer on a rotator for 5 min at RT, centrifuge, and discard supernatant.

52. Briefly wash RNA-bound beads once with 500 μL High Salt Wash buffer (resuspend only), centrifuge, and discard supernatant.

53. Wash RNA-bound beads twice with 500 μL TET buffer on a rotator for 5 min at RT, centrifuge, and discard supernatant.

54. Elute RNA by resuspending in 125 μL GRO-seq elution buffer and incubating for 10 min at 42 °C. Mix every 2–3 min. Pellet beads by centrifugation at $1000 \times g$ for 2 min, and transfer supernatant to a new 1.5 mL LoBind tube.

55. Repeat **step 19** three more times, pooling the eluates in one tube.

56. Extract the Br-UTP-enriched RNA (~500 μL) with one volume of acid phenol:chloroform, centrifuge, and transfer aqueous phase to a new 1.5 mL LoBind tube.

57. Extract with one volume of chloroform, centrifuge, and transfer aqueous phase to a new 1.5 mL LoBind tube.

58. Add 1/100th volume GlycoBlue and precipitate RNA by adding three volumes cold 100% ethanol and incubating at −20 °C for 20 min (*see* **Note 22**).

59. Dissolve recovered Br-UTP-enriched RNA in 8 μL DEPC-treated H$_2$O, and keep on ice.
60. *Reverse transcription of RNA*: To the Br-UTP-enriched, poly (A)-tailed RNA from **step 54**, add 1 μL dNTP mix (10 mM each) and 2.5 μL 12.5 μM GRO-seq RT PRIMER. Incubate for 3 min at 75 °C and chill on ice for 2 min.
61. Add 0.5 μL SUPERase-In, 3.75 μL 0.1 M DTT, 2.5 μL 25 mM MgCl$_2$, 5 μL 5× RT buffer, and 2 μL Superscript III reverse transcriptase. Incubate for 30 min at 48 °C.
62. To remove excess GRO-seq RT PRIMER, add 3.2 μL 10× Exonuclease I buffer and 4 μL Exonuclease I. Incubate for 1 h at 37 °C.
63. To remove RNA, add 1.8 μL 1 M NaOH and incubate for 20 min at 98 °C. Stop/neutralize the reaction with ~1.6 μL 1 M HCl (*see* **Note 28**).
64. Add 265 μL 1× TE for a final volume of 300 μL.
65. Extract the first-strand cDNA (300 μL) with one volume of phenol:chloroform:isoamyl alcohol pH 8, centrifuge, and transfer aqueous phase to a new 1.5 mL LoBind tube.
66. Extract with one volume of chloroform, centrifuge, and transfer aqueous phase to a new 1.5 mL LoBind tube.
67. Precipitate the first-strand cDNA by adding 1/100th volume GlycoBlue, 18 μL 5 M NaCl, and 900 μL (three volumes) cold 100% ethanol. Incubate 20 min at −20 °C (*see* **Note 22**).
68. Pellet cDNA by centrifugation at ≥15,000 × g for 10 min at 4 °C and discard supernatant.
69. Wash cDNA pellet with 1 mL 70% ethanol, centrifuge at ≥15,000 × g for 5 min, and remove as much supernatant as possible.
70. Air-dry and dissolve pellet in 30 μL nuclease-free H$_2$O.
71. *Size selection of cDNA to remove excess RT primer*: Run cDNA on a 2% gel cassette on the BluePippin with a 120–600 bp selection range (*see* **Note 29**), according to manufacturer recommendations. Transfer the eluted cDNA (~40 μL) to a new 1.5 mL LoBind tube. Wash the collection chamber with another 40 μL of electrophoresis buffer and combine. Add 220 μL nuclease-free H$_2$O to bring volume to 300 μL.
72. Extract the eluted cDNA (300 μL) with one volume of phenol:chloroform:isoamyl alcohol pH 8, centrifuge, and transfer aqueous phase to a new 1.5 mL LoBind tube.
73. Extract with one volume of chloroform, centrifuge, and transfer aqueous phase to a new 1.5 mL LoBind tube.

74. Precipitate the first-strand cDNA by adding 1/100th volume GlycoBlue, 18 μL 5 M NaCl, and 900 μL (three volumes) cold 100% ethanol. Incubate 20 min at −20 °C (*see* **Note 22**).

75. Pellet cDNA by centrifugation at ≥15,000 × *g* for 10 min at 4 °C and discard supernatant.

76. Wash cDNA pellet with 1 mL 70% ethanol, centrifuge at ≥15,000 × *g* for 5 min, and remove as much supernatant as possible.

77. Air-dry and dissolve pellet in 8 μL nuclease-free H_2O.

78. *Circularization of first-strand cDNA*: Add 1 μL CircLigase buffer, 0.5 μL 1 mM ATP, 0.5 μL 50 mM $MnCl_2$, and 0.5 μL CircLigase. Incubate for 1 h at 60 °C (*see* **Note 30**).

79. Heat-inactivate by incubating for 20 min at 80 °C.

80. *Re-linearization of first-strand cDNA*: Add 3.8 μL 4× Re-linearization supplement and 1.5 μL APE1. Incubate for 1 h at 37 °C.

81. Add 285 μL 1× TE for a final volume of 300 μL.

82. Extract the re-linearized cDNA (300 μL) with one volume of phenol:chloroform:isoamyl alcohol pH 8, centrifuge, and transfer aqueous phase to a new 1.5 mL LoBind tube.

83. Extract with one volume of chloroform, centrifuge, and transfer aqueous phase to a new 1.5 mL LoBind tube.

84. Precipitate the first-strand cDNA by adding 1/100th volume GlycoBlue, 18 μL 5 M NaCl, and 900 μL (three volumes) cold 100% ethanol. Incubate 20 min at −20 °C (*see* **Note 22**).

85. Pellet cDNA by centrifugation at ≥15,000 × *g* for 10 min at 4 °C and discard supernatant.

86. Wash cDNA pellet with 1 mL 70% ethanol, centrifuge at ≥15,000 × *g* for 5 min, and remove as much supernatant as possible.

87. Air-dry and dissolve recovered cDNA in 65 μL nuclease-free H_2O.

88. *PCR amplification test*: Dilute 5 μL of each cDNA sample 1:8 with 35 μL nuclease-free H_2O. Set up 7 25 μL reactions in 0.2 mL PCR tubes: 12.75 μL H_2O, 5 μL 5× Phusion HF buffer, 0.25 μL forward (RP1) primer (25 μM), 0.25 μL reverse (RPI1) primer (25 μM), 1.25 μL dNTPs (10 mM each), 5 μL 1:8 diluted library cDNA, 0.5 μL Phusion polymerase. Mix thoroughly. Incubate reaction(s) in a thermocycler with the following cycle parameters:

    Step 1. 98 °C for 30 s.

    Step 2. 98 °C for 10 s.

    Step 3. 62 °C for 30 s.

Step 4. 72 °C for 20 s.

Step 5. Go to step 2 12–24 times (13–15 cycles total).

Step 6. 72 °C for 10 min.

Step 7. Hold at 4 °C.

Beginning at 13 cycles, remove one of the reactions every two cycles and keep on ice.

89. Add 6 μL 5× Orange G loading buffer to each tube. Run each reaction on a 2% agarose gel with a 1 kb Plus DNA ladder. Stain with fresh 0.5 μg/mL ethidium bromide and visualize PCR products on a UV transilluminator. Determine the lowest PCR cycle number that produces visible products at ~200–300 bp. For full-scale amplification of each cDNA library, use three cycles less than this number, ideally at the lower end of recommended range 10–20 cycles (*see* **Note 31**).

90. *PCR amplification of cDNA library*: Using half of the remaining cDNA from **step 87**, set up six 25 μL reactions per sample in 0.2 mL PCR tubes: 12.75 μL H$_2$O, 5 μL 5× Phusion HF buffer, 0.25 μL forward (RP1) primer (25 μM), 0.25 μL reverse (RPI1–48, depending on multiplexing strategy) primer (25 μM), 1.25 μL dNTPs (10 mM each), 5 μL undiluted library cDNA, 0.5 μL Phusion polymerase. Mix thoroughly. Incubate reaction(s) in a thermocycler with the following cycle parameters, *using the number of cycles determined in **step 89***:

Step 1. 98 °C for 30 s.

Step 2. 98 °C for 10 s.

Step 3. 62 °C for 30 s.

Step 4. 72 °C for 20 s.

Step 5. Go to step 2 9–19 times (10–20 cycles total).

Step 6. 72 °C for 10 min.

Step 7. Hold at 4 °C.

91. Pool reactions for each sample and add 150 μL H$_2$O for a final volume of 300 μL.

92. Extract the amplified cDNA (300 μL) with one volume of phenol:chloroform:isoamyl alcohol pH 8, centrifuge, and transfer aqueous phase to a new 1.5 mL LoBind tube.

93. Extract with one volume of chloroform, centrifuge, and transfer aqueous phase to a new 1.5 mL LoBind tube.

94. Precipitate the first-strand cDNA by adding 1/100th volume GlycoBlue, 18 μL 5 M NaCl, and 900 μL (three volumes) cold 100% ethanol. Incubate 20 min at −20 °C (*see* **Note 22**).

95. Pellet cDNA by centrifugation at ≥15,000 × *g* for 10 min at 4 °C and discard supernatant.

96. Wash cDNA pellet with 1 mL 70% ethanol, centrifuge at ≥15,000 × *g* for 5 min, and remove as much supernatant as possible.
97. Air-dry and dissolve recovered library DNA in 30 μL nuclease-free H_2O.
98. *Size selection of amplified GRO-seq library DNA*: Run the amplified DNA on a 2% gel cassette on the BluePippin with a 200–600 bp selection range (*see* **Note 29**), according to manufacturer recommendations. Transfer the eluted DNA (~40 μL) to a new 1.5 mL LoBind tube. Wash the collection chamber with another 40 μL of electrophoresis buffer and combine. Add 220 μL nuclease-free H_2O to bring volume to 300 μL.
99. Extract the eluted library DNA (300 μL) with one volume of phenol:chloroform:isoamyl alcohol pH 8, centrifuge, and transfer aqueous phase to a new 1.5 mL LoBind tube.
100. Extract with one volume of chloroform, centrifuge, and transfer aqueous phase to a new 1.5 mL LoBind tube.
101. Precipitate the first-strand cDNA by adding 1/100th volume GlycoBlue, 18 μL 5 M NaCl, and 900 μL (three volumes) cold 100% ethanol. Incubate 20 min at −20 °C (*see* **Note 22**).
102. Pellet DNA by centrifugation at ≥15,000 × *g* for 10 min at 4 °C and discard supernatant.
103. Wash DNA pellet with 1 mL 70% ethanol, centrifuge at ≥15,000 × *g* for 5 min, and remove as much supernatant as possible.
104. Dissolve size-selected library DNA in 10 μL nuclease-free H_2O.
105. Quantify library DNA using a Qubit DNA assay kit.
106. If necessary, repeat **steps 90** to **105** with the remaining cDNA and combine.
107. Run ≤1 μL of each library on a Bioanalyzer instrument using a high-sensitivity DNA assay kit to quantify and assess library size distribution.
108. Store GRO-seq library DNA at −80 °C. Library DNA is ready for Illumina sequencing. We recommend aiming to obtain ~4×10^7 raw reads per sample.

3.2.4 GRO-Seq Data Analysis and Visualization

Several approaches for the analysis of GRO-seq data have been described [12, 29–32]. We provide here a brief overview of the major steps in our GRO-seq analysis workflow, rather than a detailed workflow of all commands and file manipulations.

1. Signal processing and base-calling will be carried out automatically by the Illumina sequencing system control and analysis

software. Output should be in the form of one or more FASTQ files.

2. Demultiplexing of barcoded samples sequenced together in the same lane may also occur during **step 1**. Alternatively, this can be carried out using FASTX-Toolkit, allowing the user to examine the influence of mismatch settings on sample data recovery.

3. Following demultiplexing, assess the overall quality of the sequence data for each sample, using various parameters such as quality-score distribution by read position, nucleotide composition by read position, and per-read quality-score distribution. Use FastQC and FastQ Screen to highlight potential issues that require further attention.

4. Trim or remove reads containing full or partial adapter or contaminant sequences, as well as reads with low-quality bases (Q < 10) at ends, using ea-utils fastq-mcf tool. Reads that are too short (<30 bp) after clipping, and reads with a high number of low-quality positions should also be removed.

5. Align (i.e., map) GRO-seq reads to a reference genome. For human samples, we currently use the Bowtie2 aligner with the hg19/GRCh37 reference genome index to align GRO-seq reads to the reference genome.

6. Remove reads with low mapping quality (MAPQ <10), using SAMtools.

7. Sort reads by alignment location and mark/remove duplicate reads mapping to the same location using Picard tools.

8. Quantify GRO-seq reads at the gene-level, using HTSeq-count and a custom GTF annotation file (*see* **Note 32**). Combine this quantification data to compile a matrix of values for each gene or across each treatment/condition, e.g., using R, for use by DESeq2 in **step 9**.

9. Identify genes and/or transcription units with differential GRO-seq signal, i.e., changes in RNA polymerase activity, upon p53 activation by Nutlin, using DESeq2 to analyze the gene-level quantitation data from **step 8** and comparing DMSO versus Nutlin-treated samples. DESeq2 uses a statistical model that takes into account the variability characteristics of sequencing data at different abundance levels, and provides a moderated fold-change estimate [25]. Typically, we use a false discovery rate (FDR) threshold of 0.1 for considering GRO-seq signals as differential between DMSO and Nutlin-treated samples.

10. Finally, to allow visualization of GRO-seq data as genome browser tracks, generate normalized bigWig files using deepTools.

3.3 Determining Genomic Binding Locations for p53 by Chromatin Immunoprecipitation Sequencing (ChIP-Seq)

3.3.1 Formaldehyde Cross-Linking and Whole-Cell Extract Preparation

1. Treat cells in 15 cm ø plates with DMSO or Nutlin-3a, as described in Subheading 3.1, to obtain ~5 × 10^7 cells per condition/treatment at harvest (*see* **Note 11**). Include extra plate(s) to collect parallel protein and RNA samples for quality control and validation by western blotting and qRT-PCR analysis (*see* **Note 33**).

2. Discard growth media (*see* **Note 13**), and gently replace with 19 mL of 1% formaldehyde in PBS (*see* **Note 34**). Gently swirl, and incubate for 15 min at RT.

3. Quench the formaldehyde by adding, dropwise, 1 mL 2.5 M glycine (125 mM final). Gently swirl to mix, and incubate for 5 min at RT.

4. Completely remove the formaldehyde/PBS/glycine solution by aspiration (*see* **Note 35**).

5. Keep plates on ice through **step 8**.

6. Wash cells twice with 15 mL ice-cold PBS (*see* **Note 34**).

7. Aspirate second wash completely (*see* **Note 36**).

8. To harvest cells, use 1 mL ice-cold complete RIPA (with inhibitors added, *see* **Note 8**) per plate plus 1 mL extra (e.g., 6 mL for 5 plates, *see* **Note 37**). Add the whole volume of RIPA buffer to the first plate on ice and dislodge the cells using a large cell scraper. Tilt the plate and scrape all cells into the RIPA buffer. Transfer cell suspension to next plate and repeat. Once all cells are pooled, transfer to a 15 mL conical tube.

9. Snap freeze cell suspension in liquid nitrogen and store at −80 °C until sonication.

10. Repeat **steps 2–9** for each set of plates per treatment/cell line/time point (*see* **Note 38**).

11. Thaw cell suspensions rapidly in a water bath at 30 °C and place on ice.

12. Sonicate 4–6 mL of cell suspension per 15 mL conical tube using a probe sonicator (*see* **Note 39**).

13. Transfer sonicated extract to an appropriate centrifuge tube and remove insoluble material by centrifuging at 12,000 × g for 15 min at 4 °C.

14. Transfer supernatant to a new 50 mL conical tube and keep extracts on ice.

15. Quantify extract protein concentration against a BSA standard curve (0.1–1 mg/mL). Dilute an aliquot of each extract 1:4–1:8 and add 10 μL to 300 μL of prepared BCA assay reagent in a 96-well plate (as per manufacturer recommendations). Include a diluted RIPA sample as a buffer control. Incubate for 45 min at 37 °C and read absorbance at 562 nm using a plate reader. Subtract the RIPA control absorbance from each sample value and calculate concentration in mg/mL.

16. Dilute extracts to 1 mg/mL with complete RIPA and snap freeze 1 mL aliquots in liquid nitrogen. Store aliquots at −80 °C.

17. To check fragmentation, take a 100 μL sample of extract, add 200 μL 1.5× ChIP elution buffer, and 9 μL 5 M NaCl, and follow Subheading 3.3.2, **steps 16–21** below. Run 5–10 μL on a 1% agarose gel alongside 1 kb Plus DNA ladder (*see* **Note 40**). This sample will also serve as Input control DNA (*see* **Note 41**).

3.3.2 p53 Chromatin Immunoprecipitation and DNA Purification

1. Thaw a 1 mL ChIP extract aliquot (1 mg of total protein) per sample at 30 °C and place on ice (*see* **Note 42**).

2. Allow 60 μL sheep anti-mouse Dynabeads M-280 bead suspension per sample, plus additional volume for pipetting losses. Place beads on magnetic separator for 1–5 min on ice until beads are pelleted, and remove supernatant (*see* **Note 43**).

3. Prepare beads by washing twice with 10 mL RIPA in a 15 mL conical tube, placing beads on magnetic separator for 1–5 min on ice until beads are pelleted, and removing supernatant each time. Finally, resuspend beads in RIPA buffer to original volume.

4. Preclear each sample by adding 30 μL of RIPA-washed beads from **step 3** (*see* **Note 44**). Incubate on rotator for 1–2 h at 4 °C, checking that beads are kept in suspension.

5. Block remaining beads by adding the appropriate volume of 25 mg/mL BSA to give 1 mg/mL final concentration (*see* **Note 45**). Incubate on rotator for 1–2 h at 4 °C.

6. Remove beads from precleared samples by placing on magnetic separator for 1–5 min on ice until beads are pelleted and transferring 950 μL of supernatant to a new tube.

7. Add 0.5 μg of anti-p53 DO1 antibody to each sample.

8. Add 30 μL of BSA-blocked beads, from **step 5**, to each sample (*see* **Note 44**).

9. Incubate on rotator overnight at 4 °C.

10. Place samples on magnetic separator for 1–5 min on ice until beads are pelleted and discard supernatant (*see* **Note 43**).

11. Wash beads twice with ice-cold RIPA buffer, four times with ice-cold IP Wash buffer, and twice again with ice-cold RIPA buffer. For each wash, use 1 mL buffer, resuspend beads and incubate on rotator for 5 min at 4 °C, place on magnetic separator for 1–5 min on ice until beads are pelleted, and remove supernatant.

12. Resuspend beads in 100 μL nuclease-free H_2O, add 200 μL 1.5× ChIP elution buffer, and mix to resuspend.

13. To elute immune complexes, incubate for 10 min at 65 °C, mixing periodically to ensure beads remain in suspension.

14. Place samples on magnetic separator for 1 min at RT, and transfer supernatant to fresh tubes. Add 12.5 μL 5 M NaCl (200 mM final).

15. To reverse cross-linking, incubate for 5 h at 65 °C. Periodically, shake or briefly spin tubes to prevent buildup of condensation inside the lids.

16. To digest protein, add 20 μg Proteinase K and incubate for 30 min at 45 °C.

17. Purify DNA by phenol/chloroform extraction. Add one volume (300 μL) phenol:chloroform:isoamyl alcohol and vortex 10 s. Centrifuge at $\geq 15,000 \times g$ for 10 min at 4 °C and transfer the aqueous layer to a new 1.5 mL LoBind tube (*see* **Note 46**). Add one volume chloroform and vortex 10 s. Centrifuge at $\geq 15,000 \times g$ for 10 min at 4 °C and transfer the aqueous layer to a new 1.5 mL LoBind tube (*see* **Note 46**).

18. To precipitate DNA, add 1/100th volume GlycoBlue, 30 μL (1/10th volume) 3 M sodium acetate pH 4.8, and 750 μL (~2.5 volumes) ice-cold ethanol. Incubate ≥ 20 min at -20 °C (*see* **Note 22**) and pellet DNA by centrifugation at $\geq 15,000 \times g$ for 10 min at 4 °C. Remove supernatant.

19. Wash DNA pellet with 1 mL room-temperature 70% ethanol, centrifuge at $\geq 15,000 \times g$ for 10 min at 4 °C, and remove all supernatant.

20. Dissolve DNA in 100 μL nuclease-free H_2O.

21. Assess quantity, enrichment specificity, and quality of recovered DNA using Qubit, qPCR, and/or Bioanalyzer (*see* **Note 47**).

22. Store ChIP DNA at -80 °C. ChIP DNA is ready for sequencing library construction. We have used the Illumina TRUseq ChIP-seq kit, but there are a number of suitable kits available from other manufacturers. We recommend aiming to obtain $\geq 2 \times 10^7$ raw reads per sample.

3.3.3 p53 ChIP-Seq Data Analysis and Visualization

Numerous approaches are used for the analysis of ChIP-seq data, depending on whether the factor being analyzed is expected to have broad versus narrow peaks and the exact questions being asked. Readers are referred to previous extensive coverage of this topic [14, 15, 33–37]. We provide here a brief overview of the major steps in our current ChIP-seq analysis workflow, rather than a detailed list of all commands and file manipulations.

1. Signal processing and base-calling will be carried out automatically by the Illumina sequencing system control and analysis software. Output should be in the form of one or more FASTQ files.

2. Demultiplexing of barcoded samples sequenced together in the same lane may also occur during **step 1**. Alternatively, this can be carried out using FASTX-Toolkit, allowing the user to examine the influence of mismatch settings on sample data recovery.

3. Following demultiplexing, assess the overall quality of the sequence data for each sample, using various parameters such as quality-score distribution by read position, nucleotide composition by read position, and per-read quality-score distribution. Use FastQC and FastQ Screen to highlight potential issues that require further attention. For ChIP-seq data, we also recommend assessing library complexity as per ENCODE consortium standards (https://www.encodeproject.org/data-standards/terms/#library).

4. Trim or remove reads containing full or partial adapter or contaminant sequences, as well as reads with low-quality bases ($Q < 10$) at ends, using ea-utils fastq-mcf tool. Reads that are too short (<30 bp) after clipping, and reads with a high number of low-quality positions should also be removed.

5. Align ChIP-seq reads to a reference genome. For human samples, we currently use the Bowtie2 aligner with the hg19/GRCh37 reference genome index to align ChIP-seq reads to the reference genome.

6. Remove reads with low mapping quality (MAPQ <10), using SAMtools.

7. Sort reads by alignment location and mark/remove duplicate reads mapping to the same location using Picard tools.

8. To identify p53 binding sites, peaks with significant enrichment of reads must be distinguished from background reads. We currently use the HOMER [26] and/or MACS2 [15] tools for this purpose. It is important to include sequence data from an unenriched input sample (*see* **Note 41**) in this analysis to control for copy number alterations and differences in sonication and sequencing efficiency.

9. HOMER utilizes a GTF annotation file to calculate the distance from each ChIP-seq peak (i.e., putative p53 binding site) to the nearest gene. To define a high-confidence list of genes associated with p53 binding sites, this list can be further filtered to limit the allowable distance for peak-to-gene association (*see* **Note 48**).

10. Finally, to allow visualization of ChIP-seq data as genome browser tracks, generate normalized bigWig files using deepTools.

3.4 Measuring Steady-State Transcriptome Response to p53 Activation by RNA Sequencing (RNA-Seq)

3.4.1 Cell Harvest and Total RNA Extraction for RNA-Seq

1. Treat cells in 10 cm ø plates with DMSO or Nutlin-3a, as described in Subheading 3.1, to obtain ~5×10^6 cells per condition/treatment at harvest (*see* **Note 11**). Include extra plate(s) to collect parallel protein and RNA samples for quality control and validation by western blotting and qRT-PCR analysis.
2. Discard growth media (*see* **Note 13**).
3. Keep plates on ice through **step 5**.
4. Add 1 mL ice-cold PBS to each plate and dislodge the cells using a large cell scraper. Tilt the plate and scrape all cells into the PBS. Transfer ~5×10^6 cells in suspension to a 1.5 mL tube (*see* **Note 49**).
5. Pellet cells by centrifugation at $5000 \times g$ for 2 min at 4 °C. Discard supernatant.
6. Thoroughly resuspend cells in 1 mL TRIzol reagent and place on ice (*see* **Note 50**).
7. Add 0.2 mL chloroform and shake vigorously for 15 s. Incubate for 5 min at RT.
8. Centrifuge samples at $12,000 \times g$ for 15 min at 4 °C. Without disturbing the interphase, transfer 500 µL of upper aqueous layer to a new RNase-free tube.
9. To precipitate RNA, add 0.5 mL isopropanol and mix thoroughly.
10. Pellet RNA by centrifugation at $12,000 \times g$ for 10 min at 4 °C, and carefully discard supernatant (*see* **Note 51**).
11. Wash RNA pellet by adding 1 mL 75% ethanol and centrifuging at $7500 \times g$ for 5 min at 4 °C. Carefully discard supernatant (*see* **Note 51**).
12. Air-dry the RNA pellet for 5 min at RT, and dissolve in 100 µL DEPC-treated H_2O.
13. Quantify RNA using NanoDrop or similar, and/or a Qubit RNA assay kit.
14. Assess RNA quality by diluting an aliquot of each RNA sample to 5 ng/µL, and analyzing 1 µL as Total RNA on a Bioanalyzer, using an RNA Pico assay kit according to the manufacturer's instructions. Check the RNA Integrity Number (RIN) against the recommended range for the RNA-seq library kit being used.
15. Store RNA at −80 °C. Total RNA samples are now ready for mRNA enrichment by poly(A) selection.

3.4.2 Enrichment of mRNA by Poly(A) Selection

1. If not included in library preparation, enrich for mRNA using the Dynabeads mRNA Direct micro kit or similar, as recommended by the manufacturer (*see* **Note 52**).

2. Elute mRNA using buffer and volume suitable for sequencing library construction.

3. Quantify mRNA using NanoDrop or similar, and/or a Qubit RNA assay kit.

4. Assess quality by diluting an aliquot of each RNA sample to 5 ng/μL, and analyzing 1 μL as mRNA on a 2100 Bioanalyzer, using an RNA Pico assay kit as recommended by the manufacturer. Check the estimated percentage of ribosomal RNA remaining in each sample. Poly(A) selection can be repeated if necessary.

5. Store mRNA at −80 °C. mRNA is now ready for sequencing library construction.

6. We have used the TruSeq RNA Library Preparation Kit v2 for Illumina sequencing and the Ion Total RNA-seq kit v2 for Ion Torrent sequencing, but there are a number of suitable kits available from other manufacturers. We recommend aiming to obtain ~4×10^7 raw reads per sample.

3.4.3 RNA-Seq Data Analysis and Visualization

Methods for the analysis of RNA-seq data greatly depend on the questions being asked, and have been extensively covered previously [18, 25, 38, 39]. We provide here a brief overview of the major steps in our RNA-seq analysis workflow, rather than a detailed workflow of all commands and file manipulations.

1. Signal processing and base-calling will be carried out automatically by the Illumina sequencing system control and analysis software. Output should be in the form of one or more FASTQ files.

2. Demultiplexing of barcoded samples sequenced together in the same lane may also occur during **step 1**. Alternatively, this can be carried out using FASTX-Toolkit, allowing the user to examine the influence of mismatch settings on sample data recovery.

3. Following demultiplexing, assess the overall quality of the sequence data for each sample, using various parameters such as quality-score distribution by read position, nucleotide composition by read position, and per-read quality-score distribution. Use FastQC and FastQ Screen to highlight potential issues that require further attention.

4. Trim or remove reads containing full or partial adapter or contaminant sequences, as well as reads with low-quality bases ($Q < 10$) at ends, using ea-utils fastq-mcf tool. Reads that are too short (<30 bp) after clipping, and reads with a high number of low-quality positions should also be removed.

5. Align RNA-seq reads to a reference transcriptome. For Illumina data, we currently use Tophat2 with the UCSC hg19

GTF annotation file provided in the iGenomes UCSC hg19 bundle to align RNA-seq reads to the reference transcriptome (*see* **Note 53**).

6. Remove reads with low mapping quality (MAPQ <10), using SAMtools.

7. Sort reads by alignment location and mark/remove duplicate reads mapping to the same location using Picard tools.

8. Check for sequencing coverage uniformity over housekeeping genes [40] using the RSeQC geneBody_coverage.py script with the hg19.HouseKeepingGenes.bed file. Poly(A) selection of low-quality RNA samples can lead to a strong 3′ bias in coverage.

9. Quantify RNA-seq reads at the gene-level, using HTSeq-count with the UCSC hg19 GTF annotation file provided in the iGenomes UCSC hg19 bundle. Combine this quantification data to compile a matrix of values for each gene or across each treatment/condition, e.g., using R, for use by DESeq2 in **step 10**.

10. Identify genes and/or transcription units with differential RNA-seq signal, i.e., changes in mRNA level, upon p53 activation by Nutlin, using DESeq2 to analyze the gene-level quantitation data from **step 9** and comparing DMSO versus Nutlin-treated samples. DESeq2 uses a statistical model that takes into account the variability characteristics of sequencing data at different abundance levels, and provides a moderated fold-change estimate [25]. Typically, we use a FDR threshold of 0.1 for considering genes as differentially expressed.

11. Finally, to allow visualization of RNA-seq data as genome browser tracks, generate normalized bigWig files using deepTools.

4 Notes

1. As an alternative to Nutlin-3a, a racemic mixture of Nutlin-3a (active enantiomer) and Nutlin-3b (inactive enantiomer), known as Nutlin-3R (Cayman Chemical, 10004372) is available at lower cost. If using Nutlin-3R, dissolve in DMSO to make a 1000x stock concentration at 20 mM for use at 20 μM final concentration. Store in small aliquots at −20 °C.

2. Prelubricated tubes are used to prevent loss of nuclei by adherence to tube walls.

3. The use of low DNA-binding tubes is recommended for all steps where optimal DNA or RNA recovery is required.

4. Extract yield, as measured by total protein, can vary greatly by cell type and size, and will need to be optimized per cell type. Our recommendation here is based on the HCT116 colorectal cancer cell line.

5. The wash steps require a large volume of ice-cold PBS, store at 4 °C.

6. It is important to store and dilute formaldehyde at RT to prevent polymerization. Note that formaldehyde stocks are usually 37% (w/w) in H_2O, usually containing 10–15% methanol to prevent oxidation and polymerization. We recommend purchasing in small volumes to ensure freshness.

7. We recommend filtering and storing glycine at RT. If storing at 4 °C, some precipitation may occur—heat in a water bath at $\geq$30 °C to redissolve.

8. We add the following protease, phosphatase, and histone/lysine deacetylase inhibitors to RIPA buffer: 1 μg/mL pepstatin A, 2 μg/mL aprotinin, 10 μM leupeptin, 1 mM benzamidine, 10 μg/mL trypsin inhibitor, 0.5 mM phenylmethane sulfonyl fluoride (PMSF), 10 mM sodium fluoride, 0.2 mM sodium orthovanadate (must be activated), 5 mM sodium butyrate, and 5 μM trichostatin A. Alternatively, inhibitor tablets such as cOmplete Mini or PhosSTOP (Roche/Sigma Aldrich) can be used.

9. Preferred magnetic separator racks draw beads to a small area on tube wall, allowing easy and complete removal of supernatant from bottom of tube and limiting bead loss.

10. The volume and incubation time for trypsin treatment may require optimization per cell type. Alternative cell dissociation methods that result in single cells in suspension may be used.

11. Cell number and confluence should be optimized based on cell type. We aim to have ~80% confluency at the end of treatment, as this ensures that cells are still actively cycling and will therefore respond consistently to Nutlin treatment.

12. The Nutlin treatment time for GRO-seq should be 30 min to 1 h to minimize the capture of indirect transcriptional regulation. We recommend 12 h Nutlin treatment for ChIP-seq and RNA-seq samples. This can be varied but should allow enough time for accumulation of p53 protein and/or steady-state RNA. For sensitive cell lines, treatment should be short enough that substantial cell death does not occur.

13. Dispose of Nutlin-containing solutions as required by local regulations.

14. Take care to not dislodge cells from plate. Set plates at an angle for ~20 s and remove all PBS.

15. This step is essential to rinse and nicely pellet the nuclei. The pellet will be white in color and more compact than for intact cells.
16. It is important that nuclei are resuspended homogeneously. It can be helpful to resuspend initially in 300 μL before adding the final 200 μL.
17. The trypan blue solution should be filtered through a 0.22 μm filter before use to remove any debris. Nuclei should appear blue and rounded or oval against the lighter blue background—not white like intact cells. If nuclei appear fragmented and/or large amounts of debris are present, use fewer pipetting cycles in **step 7**.
18. The number of reactions needed per condition to obtain sufficient run-on material for sequencing library preparation depends on cell type and treatment. For HCT116, MCF7, and SJSA cells we use 8 reactions per condition.
19. The reactions will be very viscous. Mix reactions thoroughly by gently pipetting ~15 times using a large orifice P200 tip before incubating at 30 °C. Ensure the viscous reaction is fully discharged from the tip by holding it against the wall of the tube near the meniscus, and depressing the plunger several times.
20. If carrying out many reactions, stagger the start times by ~30 s to allow for mixing and stopping the reactions. Reaction time may need to be optimized depending on cell type.
21. The viscosity of the run-on reactions necessitates the large volume of TRIzol reagent and splitting into two tubes. TRIzol LS reagent (ThermoFisher Scientific, 10296028) could be substituted here to reduce volume. Regardless, it is essential to ensure that any clumps are thoroughly dissolved—this will require extensive pipetting and/or vortexing.
22. Optional stopping point: leave RNA/DNA to precipitate overnight at −20 °C.
23. Take care not to over-dry RNA pellet as it will not easily dissolve.
24. This is a critical step—insufficient or excessive fragmentation will lead to poor yields. Concentration of NaOH and length of incubation need to be determined for each sample type and batch of 1 M NaOH. Hydrolysis can be tested on TRIzol-extracted RNA from an equivalent number of nuclei, by incubating with NaOH for variable amounts of time followed by neutralization and visualization after agarose gel electrophoresis. Aim for fragments of ~300 nt.
25. Check neutralization (pH ~7) with pH indicator strips on a test sample prior to starting hydrolysis, and adjust volume of 1 M Tris–HCl pH 8 accordingly.

26. Note that different batches of the anti-BrdU agarose beads display variability in immunoprecipitation efficiency and/or may be contaminated with RNases. Recovery of run-on material can be measured for a pilot run-on sample by RT-qPCR with primers specific for the primary (unspliced) transcript of any expressed gene. Optional RNase removal protocol: Prior to **steps 9** and **43**, wash beads twice with low salt wash buffer (150 mM NaCl, 20 mM Tris–HCl pH 7, 1% (v/v) triton, 0.1% (w/v) SDS, 2 mM EDTA), 2–3 times with high salt wash buffer (500 mM NaCl, 20 mM Tris–HCl pH 7, 1% (v/v) triton, 0.1% (w/v) SDS, 2 mM EDTA), and twice more with low salt wash buffer, centrifuging beads at $1000 \times g$ for 2 min at RT and discarding supernatant each time.

27. Around 20 As need to be added to provide a binding site for the GRO-seq RT primer. Conditions here should add ~60 As to 0.5 µg of 20-mer RNAs.

28. Base hydrolysis is used here to eliminate RNA because the high heat is required to inactivate Exonuclease I which can also degrade single-stranded DNA. IMPORTANT: check neutralization with pH indicator strips using equivalent volumes prior to starting hydrolysis, and adjust HCl volume accordingly.

29. This size selection step is to remove excess primers and primer dimers. Alternative size selection methods can be used here, such as gel extraction after separation by PAGE [41] or solid phase reversible immobilization, e.g., SPRIselect reagent (Beckman Coulter, B23317).

30. Ensure that condensation does not collect in lid of tube: either briefly centrifuge periodically, or carry out incubation in a thermocycler with heated lid.

31. The test sample was diluted eightfold which corresponds to three cycles. It is important to minimize the number of cycles used to amplify libraries to limit the impact of PCR bias. For multiple libraries, either amplify all samples according to the number of cycles for the least-concentrated sample, or dilute libraries to achieve similar amplification efficiency.

32. The HTSeq-count tool requires an annotation file in the GTF format that defines the regions within which to quantify read counts. Up to three-quarters of human promoters harbor transcriptionally engaged, but paused, RNA polymerase II, resulting in large peaks in GRO-seq signals at the 5′ ends of these genes. Because these peaks can obscure changes in RNA polymerase activity across gene bodies, a custom GTF annotation file that excludes the first 1000 nucleotides of each gene region must be generated. For human samples, we base this on the UCSC hg19 GTF from the iGenomes bundle.

OPTIONAL: It may also be desirable to identify active transcription units (i.e., extended contiguous regions with active RNAPII) de novo, without reference to existing annotation, and to quantify GRO-seq signal within these transcription units. Tools such as FStitch [9, 30] or groHMM [31] can be used to classify regions of contiguous GRO-seq signal.

33. For HCT116 cells, we routinely treat and harvest five 15 cm ø plates per condition. Depending on cell type, cell size, and treatment, a greater number of plates may be required.

34. Depending on cell type, confluency, and treatment, cells may lift off the plate easily and reduce the extract yield. Take care to add and withdraw solutions in a manner that minimizes cell loss (e.g., from a fixed point at one edge of the plate). Nonetheless it is also important to minimize differences in timing between plates—stagger steps as necessary.

35. Dispose of formaldehyde-containing solutions according to local regulations.

36. It is important to remove as much PBS as possible to avoid dilution of the RIPA lysis buffer—rest the plate briefly at an angle and aspirate residual PBS.

37. Harvesting cells in a lower volume of RIPA can result in protein precipitation at later steps.

38. We recommend staggering harvests for different conditions to minimize delays prior to snap freezing.

39. Sonication conditions (intensity and time) must be optimized for each cell type, volume, and specific sonicator to obtain a suitable fragment size distribution for ChIP-seq, typically 200–500 bp. We strongly recommend testing a range of sonication conditions on cross-linked chromatin to determine appropriate conditions. Avoid foaming and protein denaturation by submerging sonicator tip to appropriate depth. Alternatively, bath-based sonication systems such as the Qsonica Q800R or Diagenode Bioruptor are suitable for chromatin fragmentation. Regardless of sonication conditions and method, care must be taken to prevent the samples from overheating—where possible keep sample tubes in wet ice and allow sufficient break time (~1 min) between pulses.

40. Check for enrichment of fragments in the 200–500 bp range, larger fragments will still be present. DNA fragmentation can also be assessed using a Bioanalyzer.

41. An input DNA control sample must be generated and sequenced for each cell type and fragmentation condition. This is used by ChIP-seq peak calling software to estimate and control for non-uniform background signal (e.g., due to preferential breakage during sonication of open chromatin) and chromosomal copy number alterations.

42. If intending to check ChIP enrichment by qPCR, include one extra aliquot for blank bead or normal IgG control.

43. Care should be taken to avoid bead loss when removing supernatant. Use a P1000 pipette, not a vacuum aspirator. Increasing time and/or rotating tube 90–180° in magnetic rack can help to concentrate beads in a small area on tube wall.

44. When aliquotting beads, take care to ensure that the suspension remains homogeneously mixed to avoid variations in bead amount per sample.

45. Sheared salmon sperm DNA, which would normally be added to 0.3 mg/mL, is omitted from this blocking step to avoid contamination of the final sequencing libraries.

46. We recommend using a P200 tip to transfer the aqueous layer as this helps to avoid carryover of the solvent layer.

47. Expect to recover ~5–20 ng DNA per sample. Use ~5 μL to assess enrichment by qPCR, using primers covering a known binding site and adjacent negative control region(s). If enough material is recovered, run 5 ng DNA on a Bioanalyzer high sensitivity DNA chip to check the size distribution of recovered fragments.

48. For example, we found that most genes with increases in GRO-seq signal upon p53 activation had a p53 binding site within 2.5 kb of their transcription start site [9]. Alternatively, other approaches such as GREAT [42] allow for more sophisticated peak-to-gene association rules. It is important to note that, short of measuring three-dimensional inter-chromatin contacts, it is not possible to identify distal or non-linear peak-to-gene associations.

49. It is important not to exceed the recommended cell number as this could result in incomplete lysis and/or inactivation of enzymes. If cell yield allows, we find it helpful here to collect two approximately equal cell pellets here—one for extraction of total RNA, and one for whole-cell lysate to analyze by western blotting to verify stabilization of p53 by Nutlin treatment as well as other markers.

50. It is critical to disperse all clumps of cells to ensure complete lysis and inactivation of enzymes by the TRIzol reagent. Optional stopping point: store TRIzol-solubilized samples at −20 °C until proceeding.

51. RNA pellets may dislodge easily. If using an aspirator, we recommend leaving 50–100 μL of supernatant and removing this by hand using a P200 pipettor.

52. We recommend poly(A) selection over depletion of ribosomal RNA for most experiments to maximize sequencing reads from exons.

53. For most experiments, alignment to a reference transcriptome (i.e., known genes and transcripts) will be sufficient. However, it is also possible to run Tophat2 in de novo transcriptome mode. For Ion Torrent data, we have used the GSNAP aligner, as this allows for setting allowed mismatches as a percentage of read length which is required for variable read length data to avoid a bias against alignment of longer reads.

Acknowledgments

We would like to thank W. Lee Kraus and lab for help with development of the GRO-seq protocol, Robin Dowell and Mary Allen for help with GRO-seq analysis, and John T. Lis and Hans Salamanca for the RNase removal protocol for the anti-BrdU agarose beads. We thank Dave Murray for helpful discussion. This work was supported primarily by NIH grant R01 CA117907. Additional support was provided by NSF grant MCB-1817582 and the Howard Hughes Medical Institute through an Early Career Award to J.M.E between 2009 and 2015.

References

1. Sullivan KD, Galbraith MD, Andrysik Z et al (2018) Mechanisms of transcriptional regulation by p53. Cell Death Differ 25(1):133–143. https://doi.org/10.1038/cdd.2017.174
2. Sullivan KD, Gallant-Behm CL, Henry RE et al (2012) The p53 circuit board. Biochim Biophys Acta 1825(2):229–244. https://doi.org/10.1016/j.bbcan.2012.01.004
3. Bieging KT, Mello SS, Attardi LD (2014) Unravelling mechanisms of p53-mediated tumour suppression. Nat Rev Cancer 14(5):359–370. https://doi.org/10.1038/nrc3711
4. Fischer M (2017) Census and evaluation of p53 target genes. Oncogene 36(28):3943–3956. https://doi.org/10.1038/onc.2016.502
5. Levine AJ (2018) Reviewing the future of the P53 field. Cell Death Differ 25(1):1–2. https://doi.org/10.1038/cdd.2017.181
6. Wasylishen AR, Lozano G (2016) Attenuating the p53 pathway in human cancers: many means to the same end. Cold Spring Harb Perspect Med 6(8). https://doi.org/10.1101/cshperspect.a026211
7. Khoo KH, Verma CS, Lane DP (2014) Drugging the p53 pathway: understanding the route to clinical efficacy. Nat Rev Drug Discov 13(3):217–236. https://doi.org/10.1038/nrd4236
8. Tovar C, Rosinski J, Filipovic Z et al (2006) Small-molecule MDM2 antagonists reveal aberrant p53 signaling in cancer: implications for therapy. Proc Natl Acad Sci U S A 103(6):1888–1893. https://doi.org/10.1073/pnas.0507493103
9. Andrysik Z, Galbraith MD, Guarnieri AL et al (2017) Identification of a core TP53 transcriptional program with highly distributed tumor suppressive activity. Genome Res 27(10):1645–1657. https://doi.org/10.1101/gr.220533.117
10. Vassilev LT, Vu BT, Graves B et al (2004) In vivo activation of the p53 pathway by small-molecule antagonists of MDM2. Science 303(5659):844–848. https://doi.org/10.1126/science.1092472
11. Allen MA, Andrysik Z, Dengler VL et al (2014) Global analysis of p53-regulated transcription identifies its direct targets and unexpected regulatory mechanisms. Elife 3:e02200. https://doi.org/10.7554/eLife.02200
12. Core LJ, Waterfall JJ, Lis JT (2008) Nascent RNA sequencing reveals widespread pausing and divergent initiation at human promoters. Science 322(5909):1845–1848. https://doi.org/10.1126/science.1162228

13. Valouev A, Johnson DS, Sundquist A et al (2008) Genome-wide analysis of transcription factor binding sites based on ChIP-Seq data. Nat Methods 5(9):829–834. https://doi.org/10.1038/nmeth.1246
14. Furey TS (2012) ChIP-seq and beyond: new and improved methodologies to detect and characterize protein-DNA interactions. Nat Rev Genet 13(12):840–852. https://doi.org/10.1038/nrg3306
15. Zhang Y, Liu T, Meyer CA et al (2008) Model-based analysis of ChIP-Seq (MACS). Genome Biol 9(9):R137. https://doi.org/10.1186/gb-2008-9-9-r137
16. Pepke S, Wold B, Mortazavi A (2009) Computation for ChIP-seq and RNA-seq studies. Nat Methods 6(11 Suppl):S22–S32. https://doi.org/10.1038/nmeth.1371
17. Kharchenko PV, Tolstorukov MY, Park PJ (2008) Design and analysis of ChIP-seq experiments for DNA-binding proteins. Nat Biotechnol 26(12):1351–1359. https://doi.org/10.1038/nbt.1508
18. Conesa A, Madrigal P, Tarazona S et al (2016) A survey of best practices for RNA-seq data analysis. Genome Biol 17:13. https://doi.org/10.1186/s13059-016-0881-8
19. Langmead B, Salzberg SL (2012) Fast gapped-read alignment with bowtie 2. Nat Methods 9(4):357–359. https://doi.org/10.1038/nmeth.1923
20. Li H, Handsaker B, Wysoker A et al (2009) The sequence alignment/map format and SAMtools. Bioinformatics 25(16):2078–2079. https://doi.org/10.1093/bioinformatics/btp352
21. Dale RK, Pedersen BS, Quinlan AR (2011) Pybedtools: a flexible python library for manipulating genomic datasets and annotations. Bioinformatics 27(24):3423–3424. https://doi.org/10.1093/bioinformatics/btr539
22. Quinlan AR, Hall IM (2010) BEDTools: a flexible suite of utilities for comparing genomic features. Bioinformatics 26(6):841–842. https://doi.org/10.1093/bioinformatics/btq033
23. Anders S, Pyl PT, Huber W (2015) HTSeq--a python framework to work with high-throughput sequencing data. Bioinformatics 31(2):166–169. https://doi.org/10.1093/bioinformatics/btu638
24. Huber W, Carey VJ, Gentleman R et al (2015) Orchestrating high-throughput genomic analysis with Bioconductor. Nat Methods 12(2):115–121. https://doi.org/10.1038/nmeth.3252
25. Love MI, Huber W, Anders S (2014) Moderated estimation of fold change and dispersion for RNA-seq data with DESeq2. Genome Biol 15(12):550. https://doi.org/10.1186/s13059-014-0550-8
26. Heinz S, Benner C, Spann N et al (2010) Simple combinations of lineage-determining transcription factors prime cis-regulatory elements required for macrophage and B cell identities. Mol Cell 38(4):576–589. https://doi.org/10.1016/j.molcel.2010.05.004
27. Ramirez F, Dundar F, Diehl S et al (2014) deepTools: a flexible platform for exploring deep-sequencing data. Nucleic Acids Res 42(Web Server issue):W187–W191. https://doi.org/10.1093/nar/gku365
28. Wang L, Wang S, Li W (2012) RSeQC: quality control of RNA-seq experiments. Bioinformatics 28(16):2184–2185. https://doi.org/10.1093/bioinformatics/bts356
29. Allison KA, Kaikkonen MU, Gaasterland T et al (2014) Vespucci: a system for building annotated databases of nascent transcripts. Nucleic Acids Res 42(4):2433–2447. https://doi.org/10.1093/nar/gkt1237
30. Azofeifa JG, Allen MA, Lladser ME et al (2017) An annotation agnostic algorithm for detecting nascent RNA transcripts in GRO-Seq. IEEE/ACM Trans Comput Biol Bioinform 14(5):1070–1081. https://doi.org/10.1109/TCBB.2016.2520919
31. Chae M, Danko CG, Kraus WL (2015) groHMM: a computational tool for identifying unannotated and cell type-specific transcription units from global run-on sequencing data. BMC Bioinformatics 16:222. https://doi.org/10.1186/s12859-015-0656-3
32. Guzman C, D'Orso I (2017) CIPHER: a flexible and extensive workflow platform for integrative next-generation sequencing data analysis and genomic regulatory element prediction. BMC Bioinformatics 18(1):363. https://doi.org/10.1186/s12859-017-1770-1
33. Landt SG, Marinov GK, Kundaje A et al (2012) ChIP-seq guidelines and practices of the ENCODE and modENCODE consortia. Genome Res 22(9):1813–1831. https://doi.org/10.1101/gr.136184.111
34. de Santiago I, Carroll T (2018) Analysis of ChIP-seq data in R/Bioconductor. Methods Mol Biol 1689:195–226. https://doi.org/10.1007/978-1-4939-7380-4_17
35. Jordan-Pla A, Visa N (2018) Considerations on experimental design and data analysis of chromatin immunoprecipation experiments.

Methods Mol Biol 1689:9–28. https://doi.org/10.1007/978-1-4939-7380-4_2

36. Lerdrup M, Johansen JV, Agrawal-Singh S et al (2016) An interactive environment for agile analysis and visualization of ChIP-sequencing data. Nat Struct Mol Biol 23(4):349–357. https://doi.org/10.1038/nsmb.3180

37. Steinhauser S, Kurzawa N, Eils R et al (2016) A comprehensive comparison of tools for differential ChIP-seq analysis. Brief Bioinform 17(6):953–966. https://doi.org/10.1093/bib/bbv110

38. Kvam VM, Liu P, Si Y (2012) A comparison of statistical methods for detecting differentially expressed genes from RNA-seq data. Am J Bot 99(2):248–256. https://doi.org/10.3732/ajb.1100340

39. Trapnell C, Roberts A, Goff L et al (2012) Differential gene and transcript expression analysis of RNA-seq experiments with TopHat and cufflinks. Nat Protoc 7(3):562–578. https://doi.org/10.1038/nprot.2012.016

40. Eisenberg E, Levanon EY (2013) Human housekeeping genes, revisited. Trends Genet 29(10):569–574. https://doi.org/10.1016/j.tig.2013.05.010

41. Hah N, Danko CG, Core L et al (2011) A rapid, extensive, and transient transcriptional response to estrogen signaling in breast cancer cells. Cell 145(4):622–634. https://doi.org/10.1016/j.cell.2011.03.042

42. McLean CY, Bristor D, Hiller M et al (2010) GREAT improves functional interpretation of cis-regulatory regions. Nat Biotechnol 28(5):495–501. https://doi.org/10.1038/nbt.1630

Chapter 4

Analysis of Replication Dynamics Using the Single-Molecule DNA Fiber Spreading Assay

Stephanie Biber and Lisa Wiesmüller

Abstract

DNA replication is a fundamental process of life. Any perturbation of this process by endogenous or exogenous factors impacts on genomic stability and thereby on carcinogenesis. More recently, the replication machinery has been discovered as an interesting target for cancer therapeutic strategies. Given its high biological and clinical relevance, technologies for the analysis of DNA replication have attracted major attention. The so-called DNA fiber spreading technique is a powerful tool to directly monitor various aspects of the replication process by sequential incorporation of halogenated nucleotide analogs which later can be fluorescently stained and analyzed. This chapter outlines the use of the DNA fiber spreading technique for the analysis of replication dynamics and replication structures.

Key words DNA Replication, DNA Fiber Spreading Assay, Replication Fork Stalling, Origin firing, DNA damage response

1 Introduction

Replication is the most vulnerable process during a cell's lifetime [1]. Unrepaired DNA damage or delayed transcription intermediates [2] which manage to get into the S-phase hamper the replication process. Vice versa restrained replication causes replication slowdown or even stalling of replication forks and ultimately, upon prolonged failure to resolve these intermediates, the replication forks can collapse [3]. In this way as severe DNA damage as DNA double-strand breaks can accumulate, which induce genomic instability and promote carcinogenesis [1]. Therefore, DNA replication and the DNA damage response are two processes which are intricately linked. Consequently, understanding how cells deal with replication obstacles is getting more and more into focus. Endogenous sources can be rapid proliferation in tumor cells or the aging-related functional decline of the replication machinery in hematopoietic stem cells [4]. Exogenous sources can be genotoxic stress, such as during radio- and chemotherapy. Cells have developed an orchestra of replication stress response mechanisms to cope with these obstacles. Patients harboring genetic mutations disrupting replication stress response mechanisms are prone to develop cancer and show symptoms of premature aging. A better understanding of replication processes therefore provides insight into the mechanisms of aging, cancer development, and cancer treatment responses. Cancer cells feature rapid proliferation rates which induce replication stress [5] and replication stress response mechanisms such as the novel DNA damage bypass mechanism involving p53 and the specialized Translesion Synthesis (TLS)-Polymerase iota (POLι) [6]. So far, the most powerful yet relatively simple method to directly visualize the dynamics of the replication process on a genome-wide level at the resolution of single molecules resembles the DNA fiber spreading assay [7]. This assay directly visualizes the replication process at individual replication sites in the genome allowing to get information about replication elongation speed and replication fork stalling, the activity of exonucleases possibly resecting nascent DNA on stalled forks, replication fork terminations and initiations, i.e., origin firing [7].

The underlying principle of the DNA fiber spreading assay is the sequential incorporation of two thymidine analogs (CldU = 5-Chloro-2′deoxyuridine; IdU = 5-Iodo-2′deoxyuridine) into newly synthesized DNA. The immunostaining of both analogs allows the direct visualization of the newly incorporated nucleotides into single DNA molecules via microscopic visualization [8]. Methods used before the DNA fiber spreading assay mainly visualized replication at the cellular level, again exploiting the principle of incorporation of nucleotide analogs such as BrdU (5-Bromo-2′-deoxyuridine). These earlier approaches quantified the overall rates of DNA synthesis by flow cytometry or immunofluorescence analysis. The development of the DNA fiber spreading assay enabled scientists to gain insight into the dynamics of DNA replication by investigating the replication fork speed via track nucleotide incorporation track length measurements, replication fork stalling by analysis of fork asymmetries, origin firing by analysis of bidirectional tracks, and also the exonucleolytic degradation of nascent DNA by shortening of tracks [6, 8].

2 Materials

Culture the cells in appropriate medium and conditions, especially for the respective cells you want to use. For best results, the cells should be under exponential growth conditions. Therefore, seed adherent cells one or two days before the start of the experiment in wells of a 24-well or 6-well plate. It is recommended to split suspension culture cells in particular 1 day before the start of the experiment to ensure best growth conditions. Prepare and store all reagents at room temperature unless it is indicated otherwise.

2.1 Nucleotide Analogs

Work at low light!

1. CldU: 5-Chloro-2′deoxyuridine (Sigma-Aldrich, Steinheim, Germany or Cayman Chemical, Ann Arbor, USA); $C_9H_{11}ClN_2O_5$. Prepare 20 mM Stock Solution: Weigh 25 mg CldU and transfer to a 15 ml tube. Add DMSO to a volume of 4.25 ml. Mix properly (*see* **Note 1**). Split into 100 μl aliquots and store (protected from light) at −20 °C for up to 3 years.

2. IdU: 5-Iodo-2′deoxyuridine (Sigma-Aldrich, Steinheim, Germany); $C_9H_{11}IN_2O_5$: Prepare 200 mM Stock Solution (*see* **Note 2**): Weigh 0.625 g IdU and transfer to a 15 ml tube. Add 0.2 N NaOH (AppliChem, Darmstadt, Germany) to a volume of 8.75 ml. Rotate for approximately 2 h at 37 °C (low light). Mix properly (*see* **Note 1**) and make 100 μl aliquots and store at −20 °C for up to 3 years.

2.2 Lysis Buffer

Prepare Stock Solutions: 10% SDS (Sigma-Aldrich, Steinheim, Germany); 1 M Tris–HCl (Sigma-Aldrich, Steinheim, Germany; pH 7.4, *see* **Note 3**, store at 4 °C), 0.5 M EDTA (Sigma-Aldrich, Steinheim, Germany, store at 4 °C).

Prepare the working solution immediately prior to use!

For 100 μl of lysis buffer, mix 0.5% SDS (5 μl of stock solution), 200 mM Tris–HCl (pH 7.4, 20 μl of stock solution), 50 mM EDTA (10 μl of stock solution) in 65 μl H_2O. Store it at 4 °C or on ice until usage.

2.3 Fixation Solution

Mix methanol (Sigma-Aldrich, Steinheim, Germany) and acetic acid (Sigma-Aldrich, Steinheim, Germany) in a 3:1 volume ratio prior to use.

2.4 Storage and Cleaning Solution for Glass Slides

Prepare 70% Ethanol (Sigma-Aldrich, Steinheim, Germany) for the storage of the slides. Prepare 96% Ethanol for washing the slides immediately before spreading of the DNA to remove any fat or dirt from the surface.

2.5 Glass Slides and Coverslips

1. Coverslips (24 × 60 mm): Menzel-Gläser, Braunschweig, Germany.
2. Microscopic glass slides: cut edges, frosted end, Menzel-Gläser, Braunschweig, Germany.

2.6 Solutions for Immunofluorescence Staining

1. Blocking solution: Prepare 5% Bovine Serum Albumin (BSA, Sigma-Aldrich, Steinheim, Germany) in 1× Dulbecco's phosphate buffered saline (DPBS, Gibco, Darmstadt, Germany): Weigh 1 g BSA and transfer it to a 50 ml tube and mix with 20 ml 1× DPBS (diluted from 10× DPBS). Put it on 37 °C for 45 min or until it is completely diluted.

2. Primary antibodies: purified anti-BrdU mouse monoclonal antibody, clone B44 (BD [Becton, Dickinson and Company] BioScience, Franklin Lakes, USA) for IdU-detection and anti-BrdU rat monoclonal antibody, clone BU1/75 [ICR1] (Abcam, Cambridge, UK) for CldU-detection. Dilute anti-BrdU mouse monoclonal antibody 1:70 and anti BrdU rat monoclonal antibody 1:140 in 0.5% of the blocking solution.

3. Secondary antibodies: anti-mouse AlexaFluor555 (Invitrogen, Karlsruhe, Germany/ThermoFisher Scientific, Rockford, USA) or anti-rat AlexaFluor488 (Invitrogen, Karlsruhe, Germany/ThermoFisher Scientific, Rockford, USA).

4. Mowiol/DAPCO for mounting of the glass slides: Mix 75% (v/v) Mowiol (Calbiochem, Darmstadt, Germany) and 25% (v/v) DABCO (1,4-Diazabicyclo[2.2.2]octane, Sigma-Aldrich, Steinheim, Germany) immediately before use. Just freeze and thaw this mixture once.

5. Prepare stock solution of Mowiol: dissolve 20 g Mowiol in 80 ml 1× DPBS over night at room temperature, then add 40 ml 100% glycerol (Roth, Karlsruhe, Germany) and then let it rotate again for 24 h. Afterwards, centrifuge at $300 \times g$ for 15 min at room temperature. Make aliquots and freeze at $-20\ °C$.

6. Prepare DABCO stock solution: dissolve 4 g DABCO in 20 ml 1× DPBS and make Aliquots. Freeze aliquots at $-20\ °C$.

7. 2.5 N HCl: dilute 37% (12 N) HCl in H_2O under the hood.

8. 0.05% DPBS-Tween: Dilute 1× DPBS starting from 10× DPBS (Gibco, Darmstadt, Germany) with H_2O and add 500 μl Tween 20 (Merck, Darmstadt, Germany) to 1 l of 1× DPBS.

2.7 Image Acquisition and Analysis

1. Fluorescence microscope (e.g. Keyence BZ-9000, Keyence, Neu-Isenburg, Germany).

2. Analysis software enabling measurements of pixels and conversion into μm.

3. BZ-II Analyzer (Keyence, Neu-Isenburg, Germany).

4. Fiji64bit (https://fiji.sc/).

5. Zeiss Zen Software (Zeiss, Oberkochen, Germany).

6. LSM Image Browser (Zeiss, Oberkochen, Germany).

7. Software for graphic presentation and statistics analysis (e.g. GraphPadPrism Software, LaJolla, USA).

3 Methods

A schematic overview of the DNA fiber spreading assay is shown in Fig. 1. The text below describes the method in every detail. All steps are performed at room temperature unless it is indicated otherwise.

3.1 In Vivo Labeling of Nascent DNA: Incorporation of Halogenated Thymidine Analogs into Adherently Growing Cells

Two different kinds of halogenated thymidine analogs are sequentially incorporated into exponentially growing cells. Try to work at low light to protect halogenated nucleotides due to their light sensitivity. It is recommended to use labeling periods of 10 min to 30 min to consider the rapid polymerase activity and increased terminations due to prolonged labeling periods. Time is a very relevant factor. Therefore, work as precisely as you can.

1. First remove the medium, wash with warm 1× DPBS (pre-warm it to 37 °C) once and add CldU-containing medium with a concentration of 20 μM. Gently stir the plate and incubate for a certain period of time at 37 °C.

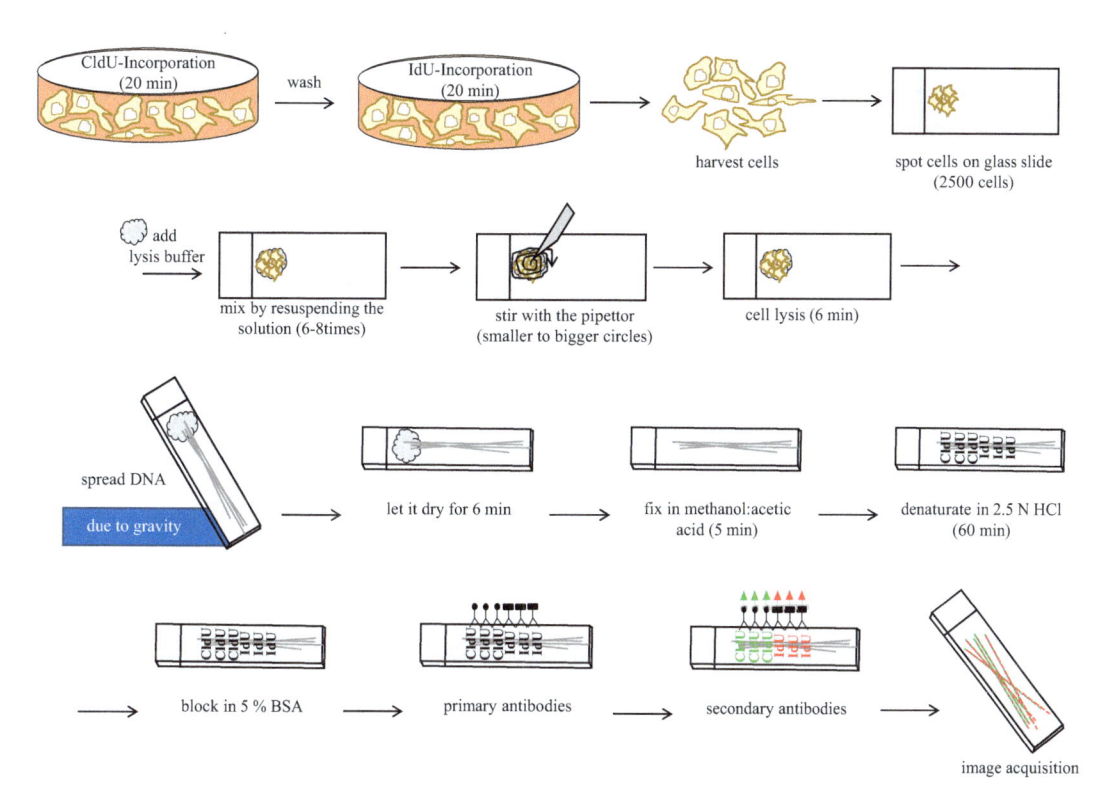

Fig. 1 Schematic overview of the DNA fiber spreading assay: Halogenated thymidine analogs are sequentially incorporated for a specific period of time. Then cells are harvested and lysed on the glass slide. Afterwards, the glass slide is tilted to an angle of about 25°–60° allowing the DNA fibers to spread down the glass slide via gravity. After fixation and denaturation, an immunostaining is performed

2. After CldU incorporation, remove the CldU-containing medium, wash with warm 1× DPBS (pre-warm it to 37 °C) twice and add IdU-containing medium with a final concentration of 200 µM. Gently stir the plate and incubate again for a certain period of time at 37 °C.

3. Afterwards, remove IdU-containing medium, wash with ice-cold 1× DPBS twice and add trypsin for cell-harvesting. Stop trypsinization immediately after dissociation of the cells from the surface with FBS-containing medium. Transfer the cell suspension into a 1.5 ml tube and centrifuge at 430 × *g* for 2 min at 4 °C (*see* **Note 4**).

4. Discard supernatant and resuspend the pellet in 1 ml of ice-cold 1× DPBS. Centrifuge at 430 × *g* for 2 min at 4 °C.

5. Discard supernatant and resuspend the pellet with ice-cold 1× DPBS approximately the same volume of the cell pellet (e.g. for a 24 well-plate approximately 20 µl is appropriate).

6. Count the cells. Keep the labeled cells on ice during determination of the cell number.

7. Adjust your cell number to a final concentration of 1250 cells per µl with ice-cold 1× DPBS (*see* **Note 5**).

3.2 In Vivo Labeling of Nascent DNA: Incorporation of Halogenated Thymidine Analogs into Suspension Culture Cells

Two different kinds of halogenated thymidine analogs are sequentially incorporated into exponentially growing cells. Therefore, split your suspension cells or add fresh medium 24 h prior to start of experiment. Alternatively, seed your cells in fresh medium and wait for 2 h until you start with the nucleotide incorporation. Try to work at low light conditions to protect halogenated nucleotides. It is recommended to use labeling periods of 10 min to 30 min to consider the high processivity of replicative polymerases and increasing frequency of terminations during prolonged labeling periods. Time is a very relevant factor. Therefore, work as precisely as you can.

1. Seed 2–4 × 10^5 cells in a 1.5 ml tube.

2. Centrifuge at 430 × *g* for 2 min. Discard supernatant and resuspend the pellet in 1 ml of CldU-containing medium with a final concentration of 20 µM. Incubate for a certain period of time at 37 °C.

3. After CldU incorporation, centrifuge at 430 × *g* for 2 min. Discard supernatant.

4. Resuspend pellet with 1 ml of IdU-containing medium with a final concentration of 200 µM. Incubate again for a certain period of time at 37 °C.

5. After IdU incorporation, centrifuge at 430 × *g* for 2 min. Discard supernatant.

6. Resuspend the pellet with ice-cold 1× DPBS and centrifuge at 430 × *g* for 2 min at 4 °C.

7. Discard supernatant and resuspend the pellet with ice-cold 1× DPBS approximately the same volume of the cell pellet.

8. Count the cells. Keep the labeled cells on ice during determination of the cell number.

9. Adjust your cell number to a final concentration of 1250 cells per μl with ice-cold 1× DPBS (*see* **Note 5**).

3.3 Spreading of the Pulse-Labeled DNA on Glass Slides

Before starting preparations for the DNA spreading procedure clean the microscopic glass slides you use for the spreading. Therefore, shortly incubate the glass slides in 96% ethanol and let them dry on paper towels. Additionally, prepare the fixation (3:1 methanol:acetic acid) and storage (70% ethanol) solution in the hood. Moreover, the lysis buffer should be prepared fresh and stored on ice until usage. Your labeled cells are kept on ice to avoid the replication process to proceed.

1. Mix your labeled cells properly and add 2 μl of the cell suspension at the top of the cleaned (and completely dried) glass slide lying horizontally flat on the lab bench.

2. Add 6 μl of the lysis buffer on top of the cell suspension droplet and pipette this mixture for 6 to 8 times up and down. Then stir with the pipette tip the cell suspension drawing a spiral starting with smaller circles. Thereby, try not to touch the glass slide with your tip and avoid air bubbles which might shear the DNA. After removing the pipette tip, you should see a viscous DNA slime indicating the lysis of the nucleus.

3. Incubate for 6 min ensuring complete lysis of cells and nuclei.

4. After the lysis, lean the glass slide on a 1.5 ml tube rack allowing the glass slide to be kept at an angle of 20° to 45°. This allows the DNA fibers to spread down the slide due to gravity (*see* **Note 6**).

5. After the cell-droplet reached the end of the glass slide, remove the rest of the droplet by putting it upright on top of a paper towel.

6. Let the glass slide dry for 6 min while lying horizontally flat and being covered with aluminum foil to protect from light. Alternatively work at low light conditions. Now the DNA fibers should be detectable along traces resembling a tree with the treetop on top of the glass slide, where the cell lysis took place, and the trunk elongating downwards to the bottom of the glass slide (*see* **Note 7**).

7. Fix the DNA fibers on the glass slide by incubating the glass slides for 5 min in the fixation solution (3:1 methanol:acetic acid) within a staining-jar.

8. Let the fixation solution evaporate from the glass slide by letting it dry for 7 min while the glass slide is lying horizontally flat and protected from light by aluminum foil.

9. Store the slides in 70% ethanol within a staining-jar at 4 °C until you perform the immunofluorescence staining.

3.4 Immunofluorescence Staining of Halogenated Thymidine Analogs Incorporated into Nascent DNA

Perform all the steps at low light conditions in staining-jars unless indicated otherwise. The incubation steps during the blocking and first and second antibodies incubation periods are performed within humidified chambers to avoid dehydration (put some wet paper towels inside the box). CldU and IdU show a 3D structure which is similar to the BrdU structure, therefore certain anti-BrdU antibodies recognize these analogs. In specific, CldU and IdU can be differentially detected by two different anti-BrdU antibodies which have been reported to specifically recognize BrdU and show cross-reactions with CldU but not IdU and to specifically recognize BrdU and show cross-reactions with IdU but not CldU, respectively [9–12].

1. Incubate glass slides in 100% methanol (Sigma-Aldrich, Steinheim, Germany) for 5 min.

2. Wash in 1× DPBS for 5 min twice.

3. Denature the DNA by incubating the glass slides in 2.5 N HCl for 1 h.

4. Wash in 1× DPBS for 5 min three times.

5. Put glass slides one by one in a humidified chamber after removing excessive 1× DPBS on the backside of the glass slide with a paper towel.

6. Blocking: add 200 μl of blocking solution (5% BSA in 1× DPBS) on glass slide and cover the glass slide with parafilm. Incubate for 45 min at 37 °C. Make sure the glass slides don't dry out (*see* **Note 8**).

7. Wash the glass slides once in 1× DPBS.

8. Primary Antibodies (also *see* **Notes 9–13**): Prepare mixture of both primary anti-BrdU antibodies (mouse and rat) diluted 1:50 in 0.5% blocking solution and let it settle for 10–20 min. Add 100 μl of antibody-mixture on the glass slide and cover it with parafilm. Incubate for 1 h (at room temperature).

9. Remove the parafilm and put the glass slides in a staining-jar filled with 0.05% 1× DPBS-Tween for washing. Wash the glass slides for 7 min during gently shaking on a shaker. Repeat washing step three times.

10. Then wash with 1× DPBS for 7 min.

11. Secondary Antibodies (light sensitive!; *see* **Note 14**): Prepare mixture of secondary antibodies diluted 1:200 (anti-mouse AlexaFluor555) and 1:400 (anti-rat AlexaFluor488) in 0.5% blocking solution. Shortly centrifuge the mixture at 2000 to 9000 × g to remove aggregates of the antibodies (*see* **Note 15**). Add 100 μl of antibody-mixture on the glass slide and cover it with parafilm. Incubate for 45 min to 1 h (at room temperature).

12. Remove the parafilm and put the glass slides in a staining-jar filled with 0.05% 1× DPBS-Tween for washing. Wash the glass slides for 10 min while gently shaking on a shaker. Repeat washing step four times.

13. Then wash with 1× DPBS for 10 min.

14. Mount the slides with 120 μl of Mowiol-Dabco mixture. Start at the bottom of the slide to put the mixture on top of the glass slide to be aware that there are no air bubbles on the top of the slide where the most important or best fibers are located. Then place the coverslips carefully on top of the glass slide and cautiously cover the complete glass slide without any air bubbles.

15. Store the slides at 4 °C in the dark until microscopy.

3.5 Image Acquisition of the DNA Fibers

The stained DNA fibers can be visualized via a fluorescence microscope. It is recommended to take pictures within regions where the DNA fibers are clearly separated, as it improves fiber analysis afterwards. It is easier to start with microscopy on the imagined treetop as most of the DNA fibers are located there. Between the big branches of DNA fibers there should be areas where the DNA fibers get more and more separated and less entangled to each other. It is easier to start with the 4× objective to get an overview and then find areas with proper DNA fibers using the 20× objective of the microscope until you take pictures with the 40× or 100× objective. Moreover, make sure that you take pictures from different areas of your slide to gain representative data. One problem of the DNA fiber spreading assay is that due to the separation of the fibers by gravity the single fibers might overlap and be entangled to each other. Therefore, it is very important to ensure that for microscopy you choose areas where the fibers are separated best.

3.6 Image Analysis of DNA Fibers

The obtained pictures can be analyzed using free software like Fiji64bit (former ImageJ, https://fiji.sc/) or Zeiss Zen Software or LSM Image Browser (both from Zeiss, Oberkochen, Germany) but also licensed software such as BZ-II Analyzer (Keyence, Neu-Isenburg, Germany). The software you use should enable you to measure the track lengths of the DNA fibers in pixel which you can then convert to μm. This allows you to calculate the replication speed or fork rate (FR) by the following formula: FR (kb/min) = [(2.59 (kb μm^{-1}) × track length (μm)/pulse time (min)]. This calculation relies on the observation that 1 μm of track length corresponds to 2.59 kb [7]. To obtain meaningful results, measure at least 150 individual fibers per experiment. For the graphical presentation of your results of the track lengths, it is not recommended to use columns because they only show the mean values but not the distribution of the track lengths of the individual fibers. Therefore, use scatter dot plots or whisker plots highlighting the distribution of the individual fiber lengths by using, e.g. GraphPadPrism Software (LaJolla, USA). To calculate the statistically significant differences between individual groups use two-tailed Mann–Whitney U test or if you need to correct for multiple comparison use Dunn's test which is a post test of the nonparametric Kruskal–Wallis test ensuring multiple comparison testing.

Besides measurement of the track lengths, this method also enables qualitative and quantitative evaluation of replication structures such as replication origins (Fig. 2c) or terminations (Fig. 2d). Therefore, you count these individual structures (best using free software like Fiji64bit) and calculate the relative frequency of the different structures as the percentage of all counted structures including or excluding ongoing forks (bi-colored, Fig. 2b). Figure 2 shows a schematic overview of the different replication structures that can be analyzed by the DNA fiber spreading assay. In general, origins represent two diverging forks, whereas terminations can be caused by two merging forks. Both origins and terminations can further be subdivided depending on the labeling period when the origin or termination event occurs (Fig. 2c, d). Notably, terminations during the first labeling period can have two explanations: either a termination event during the first labeling period whereby two forks converge to each other or stalling/termination events during or directly after the first labeling period (Fig. 2d, bottom). Techer and colleagues [13] investigated the issue whether first pulse terminations represent a true termination event or stalling events. As they hypothesized that initiation and terminations should occur at the same frequency, they measured first pulse terminations (green only) and second pulse origins (red only) and proposed that the majority of green only tracks represent true termination events. Therefore, we recommend to classify green only tracks as termination and not as stalling events.

Additionally, you can also measure replication fork stalling by calculating the fork asymmetry (Fig. 2e). Thereby, you measure the IdU-track lengths of two IdU-tracks emanating from the same origin (same CldU-track). In the case of replication fork stalling, these two IdU-tracks more frequently show different track lengths. When calculating the fork asymmetry as ratio of the longer IdU-track versus the shorter IdU-track emanating from the same

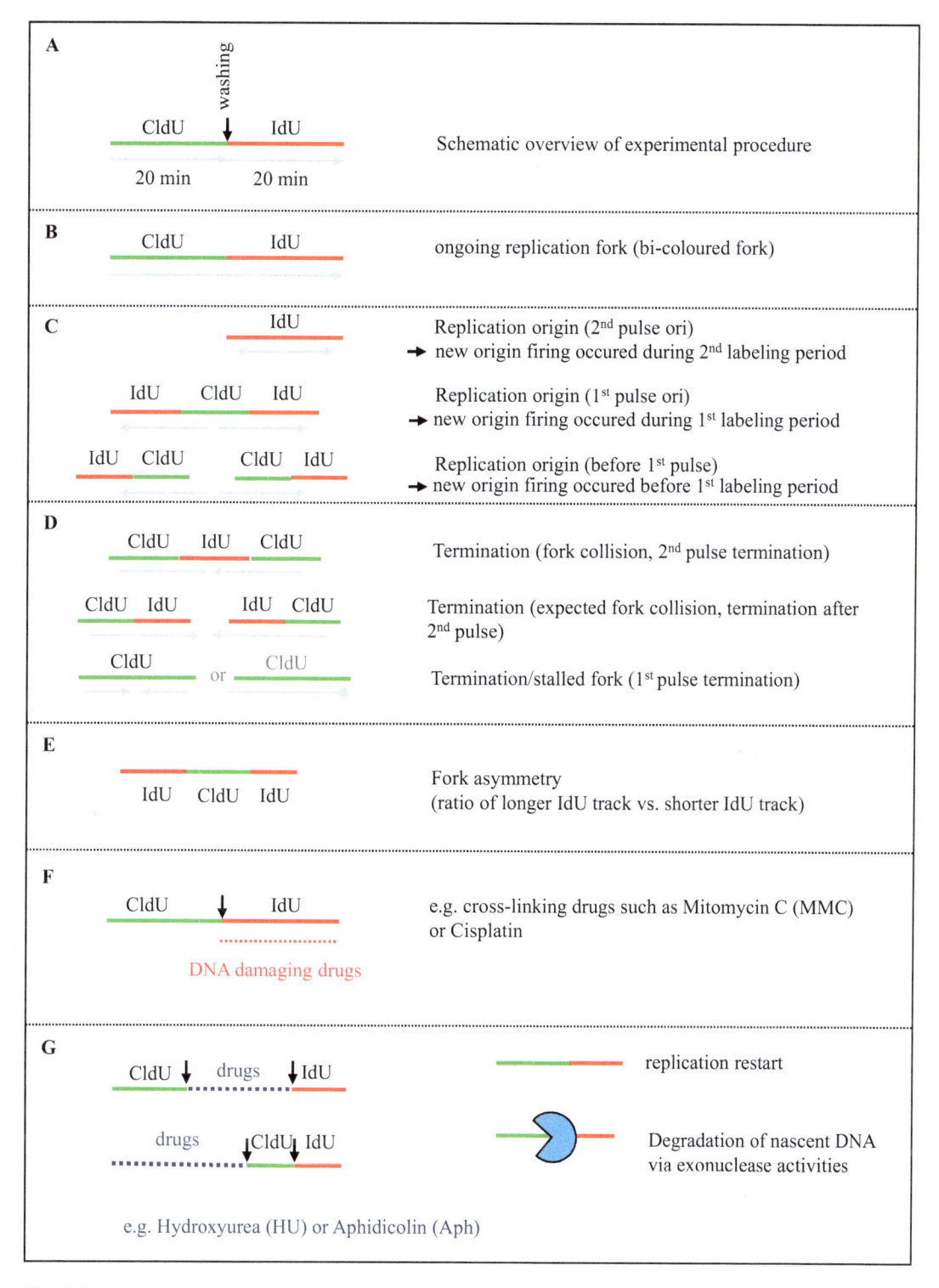

Replication Analysis Using the DNA Fiber Assay 67

Fig. 2 Schematic overview of different replication structures: (**a**) Schematic overview of the experimental procedure. CldU (green) is incorporated for 20 min. After washing (indicated by arrow throughout this figure) of the cells, IdU (red) is incorporated for another 20 min. (**b**) Scheme of an ongoing replication fork (bi-colored

CldU-track, the increase of the ratio indicates the degree of replication fork stalling (Fig. 2e). For the graphical presentation of your fork asymmetry data, you can use column bars because only the ratio but not the distribution of the ratio is important as indicator for replication fork stalling. To investigate the dynamics of replication after exogenously induced replication stress, which may cause genomic instability, you can treat the cells with replication inhibitors (Fig. 2f, g). Thereby, it is sometimes necessary to differentiate between drugs generally affecting all replication forks and drugs inducing replication barriers, hence, differentially affecting individual replication forks. Drugs inducing replication barriers throughout the whole genome at random positions are, e.g. DNA cross-linking drugs like Mitomycin C (MMC) or cisplatin (Fig. 2f). Here, you treat the cells with these drugs during the second labeling period (IdU incorporation). Treating the cells with UV-light in between the two incorporation steps is another possibility to induce replication barriers. For a general perturbation of the replication process, you treat the cells with genotoxic drugs such as Hydroxyurea (HU) or Aphidicolin (Aph). These drugs most likely and very rapidly perturb all replication forks throughout the genome in a similar manner by transiently blocking the DNA synthesis by depletion of the nucleotide pool (HU) or blocking the replicative polymerases (Aph). The treatment schedules displayed in Fig. 2g also allow you to investigate possible exonucleolytic activities acting on nascent DNA or replication fork restart after treatment and thereby replication perturbation (Fig. 2g). To investigate extensive exonucleolytic degradation of nascent DNA, you have to incorporate either one or both nucleotide analogs before start of the treatment with the genotoxic drug to perturb all replication forks. Then measure the track lengths of the labeled DNA directly before start

---

Fig. 2 (continued) fork). To obtain the track lengths for the analysis of replication speed, only the tracks of bi-colored forks are measured. (**c**) Schematic overview of replication origins. Upper panel shows new origin firing during the second labeling period. Middle panel indicates first pulse origin when new origin firing occurs during the first labeling period. Lower panel shows origin firing before incorporation of the first nucleotide CldU. (**d**) Schematic overview of different kinds of terminations. Upper panel shows the collision of two forks, called second pulse termination. The middle panel shows an expected fork collision indicating that termination might occur after the second labeling pulse. The lower panel shows first pulse termination. Theoretically this might be subdivided into real termination on the left where two forks converge during the first labeling period and into two stalling events when one fork goes into two directions and terminates or stalls during the first labeling period. (**e**) Measurement of replication fork stalling via the analysis of the fork asymmetry. The length of two IdU-tracks originating from the same CldU-track are measured and the ratio of the longer vs. shorter IdU-track is calculated. The increase of the ratio indicates replication fork stalling. (**f**) Schematic overview of the treatment with DNA damaging drugs (dotted) such as cross-linking drugs to investigate replication speed or stalling as replication forks encountering the induced replication barriers are differentially affected. (**g**) Schematic overview of the treatment to investigate replication fork restart and elongation speed after treatment with replication inhibitory drugs (dotted) affecting all replication forks

of treatment and further at different times after treatment. Due to exonucleolytic degradation, the tracks should get shorter in comparison to control cells [14]. To investigate replication fork restart you incorporate the cells with CldU, subsequently treat the cells with HU or Aph for a certain time, remove the drug by extensive washing and finally perform the IdU-labeling. You can detect replication fork restart by counting fibers labeled with both analogs (CldU—IdU; IdU—CldU—IdU, CldU—IdU—CldU) whereas stalled forks are detected by CldU-tracks only. It is recommended to perform the IdU-labeling period for only 20 min as an increase of the labeling period gives the cells more time for replication fork restart; however, this may decrease the possibility of detecting defects in replication fork restart.

4 Notes

1. If available put it on a rotator and let it rotate for 2 h at 37 °C. This allows the powder to get fully diluted.

2. The concentration of the second thymidine analog (IdU) should be tenfold higher as the concentration of the first thymidine analog (CldU). This allows that during the second labeling period the likelihood of IdU incorporation is increased in comparison to CldU which might not have fully been removed via the two washing steps between CldU and IdU incorporation.

3. Mix it properly before every usage, ensure from time to time that the pH-value is still correct, if not adjust it to the right pH-value again.

4. After cell-harvesting try to work on ice and perform centrifugation steps on 4 °C.

5. After double-labeling, harvesting, and adjusting your cells to the right concentration, you can also freeze the samples at −20 °C until you proceed with the DNA spreading.

6. If the cell solution stops to drop down the glass slide, then push the slide more upwards but be very careful not to hurtle the glass slide, this might destroy intact fibers.

7. At the treetop you should find the best fibers for microscopy. If it is hard to find appropriate positions where the fibers are clearly separated follow the treetop to the trunk and have a look at the edges of the trunk. Here, some branches along the trunk should have nice fibers for microscopy.

8. Be careful that the glass slides don't dry out. Therefore, only put the blocking solution on two or three glass slides until you take the next ones out of the $1\times$ DPBS-solution.

9. In the beginning I performed immunofluorescence staining differently: I first performed the antibody staining to detect IdU-labeled DNA (primary antibody, washing, secondary antibody, washing) and then performed the whole procedure for the CldU-labeled DNA afterwards. Sometimes, this procedure decreases the background of the immunofluorescence staining; however, this procedure is more time consuming.

10. After blocking I prepare 1:50% anti-mouse anti-BrdU antibody dilution in 0.5% blocking solution and put 100 μl on the glass slides before covering with parafilm. After 1 h incubation at room temperature, remove parafilm and proceed to the washing step as described above.

11. After washing prepare anti-mouse AlexaFluor555 secondary antibody diluted 1:200 in 0.5% blocking solution and put 100 μl on the glass slides before covering with parafilm. After 1 h incubation at room temperature, remove parafilm and proceed to the washing step as described above.

12. Then you perform the same steps again for detection of CldU: prepare a 1:70 dilution of anti-rat anti-BrdU antibody in 0.5% blocking solution, incubate for 1 h at room temperature, perform washing, prepare anti-rat AlexaFluor488 secondary antibody diluted 1:500 in 0.5% blocking solution, incubate for 45 min at room temperature, perform washing.

13. Mount the glass slides with Mowiol-Dabco mixture as described above.

14. Fluorochrome conjugated secondary antibodies are very light sensitive. Therefore, always work at low light conditions while performing the secondary antibody incubation step and the steps afterwards during the immunofluorescence staining.

15. If you have trouble with high background, it is recommended to centrifuge your secondary antibodies shortly in a microcentrifuge at highest speed to remove any aggregates. Centrifuging the mixture at 2000 to 9000 × g for 1 min is enough to remove the aggregates of the antibodies which might cause background.

Acknowledgments

We cordially thank Dr. Vanesa Gottifredi as well as Dr. Sabrina F. Mansilla for teaching Stephanie Biber the DNA fiber spreading assay during an internship at the Cell Cycle and Genomic Stability Laboratory, Fundación Instituto Leloir-IIBBA, CONICET Buenos Aires, Argentina financed by the DFG Graduate School for Molecular Medicine at Ulm University (IGradU), during her PhD fellowship funded by the IGradU. We also thank Dr. Helmut Pospiech (University of Oulu, Finland and Leibnitz

Institute on Aging, Jena, Germany) for general advice regarding the DNA fiber spreading assay.

This work was supported by the German Research Foundation (DFG, PA3 in Research Training Group 1789 "Cellular and Molecular Mechanisms in Aging," CEMMA) and by the German Cancer Aid (Deutsche Krebshilfe #70112504).

References

1. Aguilera & García-Muse (2013) Causes of genome instability. Annu Rev Genet 47:1–32
2. Sarni D, Kerem B (2017) Oncogene-induced replication stress drives genome instability and tumorigenesis. Int J Mol Sci 18(7):1339
3. Bournique E, DallÒsto M, Hoffmann JS, Bergoglio V (2018) Role of specialized DNA polymerases in the limitation of replicative stress and DNA damage transmission. Mutat Res 808:62–73
4. Flach J, Milyavsky M (2018) Replication stress in hematopoietic stem cells in mouse and man. Mutat Res 808:74–82
5. Gaillard H, Garcia-Muse T, Aguilera A (2015) Replication stress and cancer. Nat Rev Cancer 15(5):276–289
6. Hampp S, Kiessling T, Buechle K, Mansilla SF, Thomale J, Rall M, Ahn J, Pospiech H, Gottifredi V, Wiesmüller L (2016) DNA damage tolerance pathway involving DNA polymerase ι and the tumor suppressor p53 regulates DNA replication fork progression. Proc Natl Acad Sci U S A 113(30):4311–4319
7. Jackson DA, Pombo A (1998) Replicon clusters are stable units of chromosome structure: evidence that nuclear organization contributes to the efficient activation and propagation of s phase in human cells. J Cell Biol 140(6):1285–1295
8. Nieminuszczy J, Schwab RA, Niedzwiedz W (2016) The DNA fibre technique – tracking helicases at work. Methods 108:92–98
9. Gratzner HG (1982) Monoclonal antibody to 5-bromo and 5-iododeoxyuridine: a new reagent for detection of DNA replication. Science 218:474–475
10. Gray JW (1985) Monoclonal antibodies against bromodeoxyuridine (special issue). Cytometry 6:501–673
11. Aten JA, Bakker PJ, Stap J, Boschman GA, Veenhof CH (1992) DNA double labeling with IdUrd and CldUrd for spatial and temporal analysis of cell proliferation and DNA replication. Histochem J 25(5):251–259
12. Bugler B, Schmitt E, Aressy B, Ducommun B (2010) Unscheduled expression of CDC25B in S-phase leads to replicative stress and DNA damage. Mol Cancer 4:9–29
13. Tècher H, Koundrioukoff S, Azar D, Wilhelm T, Carignon S, Brison O, Debatisse M, Le Tallec B (2013) Replication dynamics: biases and robustness of DNA fiber analysis. J Mol Biol 425:4845–4855
14. Schlacher K, Christ N, Siaud N, Egashira A, Wu H, Jasin M (2011) Double-strand break repair-independent role for BRCA2 in blocking stalled replication fork degradation by MRE11. Cell 145(4):529–542

Chapter 5

Measuring Translation Efficiency by RNA Immunoprecipitation of Translation Initiation Factors

Chris Lucchesi, Shakur Mohibi, and Xinbin Chen

Abstract

Eukaryotic mRNAs are bound by a multitude of RNA binding proteins (RBPs) that control their localization, transport, and translation. Measuring the rate of translation of mRNAs is critical for understanding the factors and pathways involved in gene expression. In this chapter, we present a method to compare the rate of translation of individual mRNA species based on the fraction of mRNA bound by a specific ribonucleoprotein involved in the general translation machinery. The ribonucleoprotein complex is immunoprecipitated using an antibody for the RBP, followed by RT-PCR to semi-quantitatively determine the amount of an individual mRNA fraction bound by a translation regulating protein such as eIF4E.

Key words RNA immunoprecipitation, Translation, eIF4E

1 Introduction

A majority of eukaryotic genes are regulated at transcriptional and post-transcriptional levels giving each one a unique expression profile under specific conditions [1, 2]. Quantification of transcriptional gene expression at a global level can be achieved by high-throughput techniques, such as microarrays and RNA-seq [3]. However, as eukaryotic translation is uncoupled to transcription, the gene expression patterns often poorly correlate with the amount of protein produced for the particular gene [1]. As the final effector molecule for most of the eukaryotic genes is protein, estimation of the protein production by mRNA translation is critical in understanding important cellular pathways.

The post-transcriptional regulation of mRNAs is achieved by RNA-binding proteins (RBPs) that control the mRNA fate by determining mRNA stability, transport, localization, and eventually translation [4]. Several RBPs coordinately regulate the post-

---

Chris Lucchesi and Shakur Mohibi contributed equally to this work.

transcriptional fate of each mRNA by binding to them and forming ribonucleoprotein (RNP) complexes. Distinct groups of RBPs regulate different aspects of mRNA fate such as alternative splicing, nuclear to cytoplasmic shuttling, mRNA stability, and translation [5]. Translation initiation for the majority of eukaryotic mRNAs involves binding of the 5′ cap by RBPs of the general translation initiation machinery, such as eIF4E [6]. Thus, measuring the binding of these RBPs to the mRNA of interest can provide a way of measuring the translation efficiency of that particular mRNA.

Various techniques have been used to measure the rate of translation of an individual gene or for the entire translatome such as polysome profiling and labeling with ^{35}S-methionine or cysteine [7–10]. Polysome profiling has been extensively used for measuring translation rates and involves fractionation of polysomes bound to mRNAs from cells followed by RT-PCR for a specific mRNA or high-throughput sequencing to get the landscape of translatome in a given cell type [8–10]. On the other hand, metabolic pulse labeling with ^{35}S-methionine is followed by immunoprecipitation of a specific protein, running on SDS-PAGE and autoradiography [7, 11]. Although polysome profiling gives a better idea about mRNA translation rates, it is a labor intensive technique, whereas pulse labeling with ^{35}S requires good quality antibodies for the protein of interest.

Our laboratory uses RNA immunoprecipitation (RNA IP) of RBPs involved in translation initiation, such as the cap binding protein eIF4E, followed by quantification of the mRNA of interest bound by eIF4E [12–15]. We follow the RNA IP with semi-quantitative RT-PCR to quantify the fraction of mRNA bound by eIF4E [12–15] (Fig. 1). This provides a coarse estimation of the translation initiation of the mRNA under study in the specific condition. This technique when used concurrent with either polysome profiling or ^{35}S labeling enables for a reliable estimation of the translation rate of the gene of interest. Here we describe the RNA IP protocol followed by RT-PCR for semi-quantitatively measuring the translation efficiency of mRNAs.

2 Materials

It is recommended that all reagents, glassware and plasticware should either be certified RNase and DNase free or treated with 0.1% DEPC.

1. Yeast tRNA (1 mg ml^{-1}).
2. 3 M sodium acetate.
3. TRIzol or phenol–chloroform–isoamyl alcohol mixture.
4. 10% SDS solution.

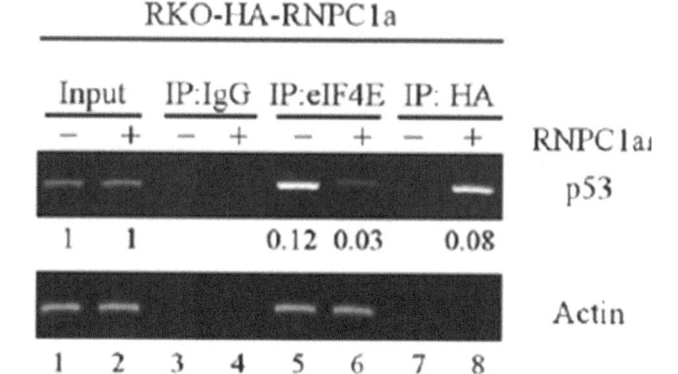

Fig. 1 Determining translation efficiency by measuring eIF4E bound to mRNA. eIF4E bound mRNAs were immunoprecipitated followed by RT-PCR with p53 mRNA specific primers to show that RNPC1a (RBM38) inhibits the translation of p53 mRNA by preventing eIF4E binding to p53 mRNA. Actin mRNA bound by eIF4E was used as a control. (Reproduced from *Genes and Development (2011) 25:1528–1543* with permission from *Cold Spring Harbor Laboratory Press*)

5. Sterile PBS.
6. 6 M Urea.
7. Proteinase K (30 μg per 100 μl).
8. Protein A or protein G agarose beads.
9. Ethanol.
10. RevertAid RT Reverse Transcription Kit.
11. DreamTaq Green PCR Master Mix.
12. Primers for gene of interest and control.
13. Antibodies for translation initiation factors (e.g., eIF4E).
14. Nuclease-free water should be used to prepare the following buffer: *Polysome lysis buffer*: 100 mM KCl, 5 mM MgCl$_2$, 10 mM HEPES, pH 7.0, 0.5% Nonidet P-40, 1 mM DTT, 100 U ml^{-1} RiboLock RNase inhibitor (Thermo, cat. no. EO0381), 2 mM vanadyl ribonucleoside complexes solution (Sigma cat. No. 94742), 25 μl ml^{-1} protease inhibitor cocktail (Sigma cat. no. P8340). Typically, we prepare a 10× stock solution of this buffer without DTT, RNase inhibitor or protease inhibitor cocktail. We prepare fresh complete polysome lysis buffer the day that it is needed by making 1× solution and adding 1 mM DTT, 2 mM vanadyl ribonucleoside complex solution, 100 U ml^{-1} RiboLock RNase inhibitor, and protease inhibitor cocktail.

3 Methods

The method used here has been successfully used for mammalian cancer cells to determine the amount of target mRNA bound to our protein of interest. We would assume that this method could be adapted and/or modified for samples from other cells or tissue. This method was modified from [16].

3.1 Lysate Preparation

Day 1: (Timing 3+ hours)

1. Harvest cells with freshly made ice cold polysome lysis buffer. For each immunoprecipitation use one 10-cm plate containing ~4.8×10^6 cells. Wash cells two times with 5 ml ice cold PBS. Add 1 ml of polysome lysis buffer per 10-cm plate. Scrape the cells with a cell scraper and transfer lysate to a 1.5 ml tube on ice.

2. Spin down at full speed ($16,000 \times g$) for 15 min in a microcentrifuge at 4 °C.

3. Transfer supernatant to a new 1.5 ml tube on ice.

4. Equilibrate 50% protein A-agarose beads in polysome lysis buffer. Wash beads twice with 0.5 ml polysome lysis buffer and then restore the original volume with lysis buffer (50% protein A-agarose slurry). 50 µl of 50% agarose slurry is needed for ever 1 ml of lysate for every preclearing step and the immunoprecipitation step. For instance, for every 4 ml of lysate you will need to equilibrate three 200-µl aliquots of slurry. Two of these 200-µl aliquots will be used for preclearing, and the remaining 200-µl aliquot will be divided into four 50 µl aliquots for the immunoprecipitation (*see* **Note 1**).

5. For preclearing, add 50 µl of equilibrated protein A-agarose beads (50% slurry) to each 1 ml of lysate and incubate rotating at 4 °C for 1 h. Centrifuge briefly to collect beads. Transfer supernatant to new 1.5 ml tube.

6. Repeat **step 5**.

7. Pool supernatant for each group, and then divide lysate into 1 ml aliquots in new 1.5 ml tubes. Take 100 µl from each group for an input control and place at −20 °C until later.

8. Add antibodies to 1 ml aliquots and rotate overnight at 4 °C. Include an antibody for a translation initiation factor such as eIF4E as well as a negative control antibody of the same type as antibody of interest (e.g., Mouse IgG). For our application we add 1 µg of antibody per sample (*see* **Note 2**).

3.2 Immunoprecipitation

Day 2: (Timing 6+ hours)

1. Add 50 μl equilibrated protein A-agarose beads to each sample and rotate at 4 °C for 4 h.
2. Centrifuge briefly to collect beads and carefully discard the lysate.
3. Wash beads with 0.5 ml polysome lysis buffer four times rotating at 4 °C for 5 min each wash. For washes, polysome lysis buffer without RNase inhibitor and protease inhibitors can be used.
4. Wash beads four times with polysome lysis buffer containing 1 M urea, as above.
5. Resuspend beads in 100 μl polysome lysis buffer with 0.1% SDS and 30 μg proteinase K. Incubate in a heating block at 50 °C for 30 min. In addition, also add 0.1% SDS and 30 μg proteinase K to your input samples that you took the day before and stored at −20 °C.
6. Add 100 μl TRIzol to each sample and vortex to mix. Centrifuge (16,000 × g) for 1 min to sperate phases. Recover the upper water phase and transfer to a new 1.5 ml tube (*see* **Note 3**).
7. Add 10 μl yeast tRNA (1 mg ml^{-1}), 12 μl 3 M sodium acetate, and 250 μl ethanol to the 100 μl water phase and mix. Continue to ethanol-precipitate overnight at −80 °C (*see* **Notes 4 and 5**).

3.3 RT-PCR

Day 3: (Timing 8+ hours)

1. Spin down at full speed (16,000 × g) for 20 min in a microcentrifuge at 4 °C.
2. Remove ethanol carefully and let the pellet air-dry until all liquid has evaporated.
3. Add 10 μl H$_2$O to the RNA pellet. Next, treat with DNase I as follows:

Sample Prep.		PCR Cycle
RNA	5 μl	37 °C—30 min
5× RT buffer	2 μl	75 °C—10 min
DNase I	1 μl	4 °C—5 min
RNase Inhibitor	0.2 μl	
H$_2$O	1.8 μl	
Total volume	10 μl	

4. After completing DNase I treatment of your RNA, continue with cDNA synthesis as follows:

Sample Prep.		PCR Cycle
RNA	4 μl	25 °C—5 min
Random Primer	1 μl	42 °C—60 min
H$_2$O	7 μl	70 °C—5 min
Mix		4 °C—10 min
5× Reaction Buffer	4 μl	
RiboLock RNase Inh.	1 μl	
10 mM dNTP	2 μl	
RevertAid RT	1 μl	
Total volume	20 μl	

5. Proceed with PCR reaction. Set up 20 μl PCR reaction using 1 μl from cDNA synthesis as template. Run PCR for 25–45 cycles. Include primers for your gene of interest as well as an internal loading control (*see* **Note 6**).

6. Visualize by running 15 μl of each reaction on an agarose gel.

4 Notes

1. The beads here depend on the particular antibody you are going to be using. For example, mouse monoclonal antibodies typically bind stronger to protein G beads, whereas rabbit polyclonal antibodies typically bind equally well to either protein A or protein G beads. It is recommended to check with the manufacturer to establish the best beads for the isotype of antibody you will be using.

2. The amount of antibody to be used will vary from antibody to antibody and will need to be determined by the scientist. In addition, not every antibody will work for immunoprecipitations.

3. Instead of TRIzol, phenol–chloroform–isoamyl alcohol mixture can be used.

4. Other carriers can be used such as acrylamide or glycogen.

5. Do not cut down on precipitation time as this may cause loss of material. You CAN increase the precipitation time if needed.

6. The amount of template as well as cycles will have to be determined by the scientist to achieve the optimal PCR product.

References

1. Orphanides G, Reinberg D (2002) A unified theory of gene expression. Cell 108(4):439–451
2. Chen K, Rajewsky N (2007) The evolution of gene regulation by transcription factors and microRNAs. Nat Rev Genet 8(2):93–103. https://doi.org/10.1038/nrg1990
3. Wang Z, Gerstein M, Snyder M (2009) RNA-Seq: a revolutionary tool for transcriptomics. Nat Rev Genet 10(1):57–63. https://doi.org/10.1038/nrg2484
4. Keene JD (2007) RNA regulons: coordination of post-transcriptional events. Nat Rev Genet 8(7):533–543. https://doi.org/10.1038/nrg2111
5. Zhang J, Chen X (2008) Posttranscriptional regulation of p53 and its targets by RNA-binding proteins. Curr Mol Med 8(8):845–849
6. Richter JD, Sonenberg N (2005) Regulation of cap-dependent translation by eIF4E inhibitory proteins. Nature 433(7025):477–480. https://doi.org/10.1038/nature03205
7. Pollard JW (1994) The in vivo isotopic labeling of proteins for polyacrylamide gel electrophoresis. Methods Mol Biol 32:67–72. https://doi.org/10.1385/0-89603-268-X:67
8. Chasse H, Boulben S, Costache V, Cormier P, Morales J (2017) Analysis of translation using polysome profiling. Nucleic Acids Res 45(3):e15. https://doi.org/10.1093/nar/gkw907
9. Kudla M, Karginov FV (2016) Measuring mRNA translation by polysome profiling. Methods Mol Biol 1421:127–135. https://doi.org/10.1007/978-1-4939-3591-8_11
10. Zuccotti P, Modelska A (2016) Studying the Translatome with Polysome profiling. Methods Mol Biol 1358:59–69. https://doi.org/10.1007/978-1-4939-3067-8_4
11. Cho SJ, Rossi A, Jung YS, Yan W, Liu G, Zhang J, Zhang M, Chen X (2013) Ninjurin1, a target of p53, regulates p53 expression and p53-dependent cell survival, senescence, and radiation-induced mortality. Proc Natl Acad Sci U S A 110(23):9362–9367. https://doi.org/10.1073/pnas.1221242110
12. Yang HJ, Zhang J, Yan W, Cho SJ, Lucchesi C, Chen M, Huang EC, Scoumanne A, Zhang W, Chen X (2017) Ninjurin 1 has two opposing functions in tumorigenesis in a p53-dependent manner. Proc Natl Acad Sci U S A 114(43):11500–11505. https://doi.org/10.1073/pnas.1711814114
13. Zhang J, Cho SJ, Shu L, Yan W, Guerrero T, Kent M, Skorupski K, Chen H, Chen X (2011) Translational repression of p53 by RNPC1, a p53 target overexpressed in lymphomas. Genes Dev 25(14):1528–1543. https://doi.org/10.1101/gad.2069311
14. Zhang Y, Qian Y, Zhang J, Yan W, Jung YS, Chen M, Huang E, Lloyd K, Duan Y, Wang J, Liu G, Chen X (2017) Ferredoxin reductase is critical for p53-dependent tumor suppression via iron regulatory protein 2. Genes Dev 31(12):1243–1256. https://doi.org/10.1101/gad.299388.117
15. Zhang M, Zhang Y, Xu E, Mohibi S, de Anda DM, Jiang Y, Zhang J, Chen X (2018) Rbm24, a target of p53, is necessary for proper expression of p53 and heart development. Cell Death Differ 25(6):1118–1130. https://doi.org/10.1038/s41418-017-0029-8
16. Peritz T, Zeng F, Kannanayakal TJ, Kilk K, Eiriksdottir E, Langel U, Eberwine J (2006) Immunoprecipitation of mRNA-protein complexes. Nat Protoc 1(2):577–580. https://doi.org/10.1038/nprot.2006.82

Chapter 6

DNA Affinity Purification: A Pulldown Assay for Identifying and Analyzing Proteins Binding to Nucleic Acids

Gerd A. Müller and Kurt Engeland

Abstract

The interaction of proteins with DNA plays a central role in gene regulation. We describe a DNA affinity purification method that allows for identification and analysis of protein complex components. For example, a DNA probe carrying a transcription factor binding site is used to purify proteins from a nuclear extract. The proteins binding to the probe are then identified by mass spectrometry. In similar experiments, proteins purified by this pulldown method can be analyzed by Western blot. Employing this method, we found that the DREAM transcriptional repressor complex binds to CHR transcriptional elements in promoters of cell cycle genes. This complex is important for cell cycle–dependent repression and as part of the p53-DREAM pathway serves as a link for indirect transcriptional repression of target genes by the tumor suppressor p53. In general, the methods described can be applied for the identification and analysis of proteins binding to DNA.

Key words DNA-binding protein complexes, Transcription factors, Promoter elements, Transcription factor binding sites, DNA affinity purification, DNA–protein binding, Phylogenetic footprinting, Stable isotope labeling with amino acids in cell culture—SILAC, Mass spectrometry, Western blot

1 Introduction

Proteins binding to DNA have long been of interest in basic research. General chromatin-associated proteins such as histones and in particular their various modifications have been studied to understand regulation of chromatin accessibility and condensation. Furthermore, DNA binding of transcription factors is central to gene regulation. Their binding to specific recognition sites in promoters as well as modulation of this DNA association controls gene expression. Revolutionary improvements in two important methods have enabled us to trace also minute protein quantities of such factors in vitro as well as in vivo.

As one method, mass spectrometry was developed into a technique capable of detecting large molecules such as proteins in vitro with impressive sensitivity. Another method that was developed during the last two decades and that permits to estimate DNA binding of proteins in living cells is chromatin immunoprecipitation—ChIP.

Both methods were at the center of identifying and studying the proteins that bind a gene regulatory element that we had named *cell cycle genes homology region*—CHR [1, 2]. The CHR controls the expression of many cell cycle genes and can act as a single element or can function as part of a tandem site with the *cell cycle–dependent element*—CDE—four nucleotides upstream of the CHR. For a long time, we and others searched for the transcription factors binding to this site [3].

To identify the regulating protein complexes, we employed a method that can be generally applied to purify and identify proteins that bind to nucleic acids [4]. We used two DNA probes that were labeled with biotin at one end to allow for purification by binding to immobilized streptavidin. The two probes only differed in that one represented the wild-type promoter fragment carrying the CDE/CHR tandem site and the other was the same fragment except that it was mutated in the CDE/CHR sites [4]. As a source for candidate proteins, nuclear extracts from cells labeled by the *stable isotope labeling with amino acids in cell culture* (SILAC) protocol were employed. The two probes were then incubated with the nuclear extracts. After affinity purification on streptavidin beads, the samples were combined and analyzed by SDS-PAGE followed by nanoLC-MS/MS [4]. In general, this method can be applied without SILAC labeling of proteins in a label-free setting.

With this method, we identified a protein complex named DREAM to bind to CHR sites [4]. For the identification of DREAM, density-arrested cells in the G0 phase of the cell cycle were used. In these resting and in early-G1 cells, DREAM binds to the CHR element and downregulates a large number of mostly cell cycle–regulating genes [4, 5]. In other phases of the cell cycle, DREAM changes its composition by releasing E2F/RB-related components from the complex and adding B-MYB or FOXM1 to form complexes called MMB or FOXM1-MuvB, respectively [6, 7]. These two protein complexes are—in contrast to the repressing DREAM—activating transcription factor complexes. However, they also bind and act through CHR sites. Thus, due to this switch in protein complexes, CHR elements change from repressing to activating sites [7, 8]. Furthermore, DREAM is also formed following the activation of the tumor suppressor p53 [9]. p53 activates the p53-p21-DREAM pathway which results in the downregulation of a large number of cell cycle genes and which ultimately contributes to cell cycle arrest [10, 11].

Following the initial identification of the binding of CHR sites by mass spectrometry, we confirmed this observation by Western blot of samples from the DNA affinity purification. With the help of specific antibodies directed against the subunits of DREAM, we verified that these proteins indeed can bind to CHR elements in vitro [4]. In order to validate DREAM binding in vivo, we employed ChIP assays. General binding of DREAM components was observed when we ChIP-assayed the chromosomal region in which CDE/CHR sites had been described. In order to test whether DREAM binding indeed requires CDE/CHR sites, we used cells stably transfected with wild-type or CDE/CHR mutant promoter constructs. We observed that only the wild-type transgenes were able to bind DREAM indicating that also in vivo binding requires CDE/CHR elements [4]. In general, DNA–protein interactions identified in DNA affinity purifications should be verified by ChIP.

Besides the methods already mentioned, a few other approaches are helpful when searching for transcription factor binding sites and the proteins binding to them. In order to identify potential binding sites in a promoter that has not yet been subject to analyses by promoter reporter assays, in silico phylogenetic footprinting can give a first indication where relevant sites can be found [12]. Usually evolutionary conservation of DNA elements indicates functional relevance. The *Cyclin B2* promoter may serve as an example (Fig. 1a). Phylogenetic analysis showed that CDE and CHR elements are highly conserved. Also three CCAAT boxes often found in cell cycle promoters are detected in this inspection as phylogenetically conserved (Fig. 1a). CCAAT boxes can bind three subunits of the transcription factor NF-Y (Fig. 1b) [3]. The binding of the subunit NF-YA to CHR-regulated promoters also gives an experimental advantage as this protein can—in addition to histones—serve as a positive control for protein complex binding. After DNA affinity purification and Western blot, NF-YA can be detected also by binding to mutant probes because it binds independently of the CDE/CHR site (Fig. 1c).

Taken together, these methods allow for the identification of relevant protein binding sites, the identification of protein complexes binding to DNA and the verification of such binding in vitro and in cells. Here, we describe how DNA affinity purification can be applied to identify and analyze protein interactions with specific promoter elements.

2 Materials

2.1 Nuclei Extraction Buffer

Nuclei Extraction Buffer: 10 mM HEPES–KOH, 1.5 mM $MgCl_2$, 10 mM KCl, pH 7.9. To prepare 500 ml of the Nuclei Extraction Buffer, weigh 1.19 g HEPES, 0.15 g $MgCl_2 \cdot 6H_2O$, and 0.37 g KCl. Transfer to a 500 ml graduated cylinder and add 400 ml ultrapure water. Mix, adjust pH with KOH, and add water to 500 ml. Store at 4 °C or at −20 °C.

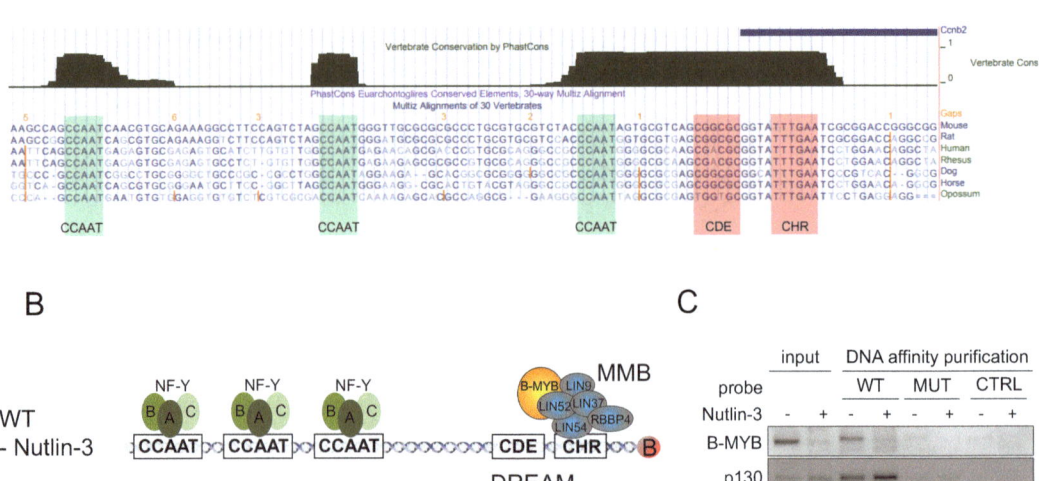

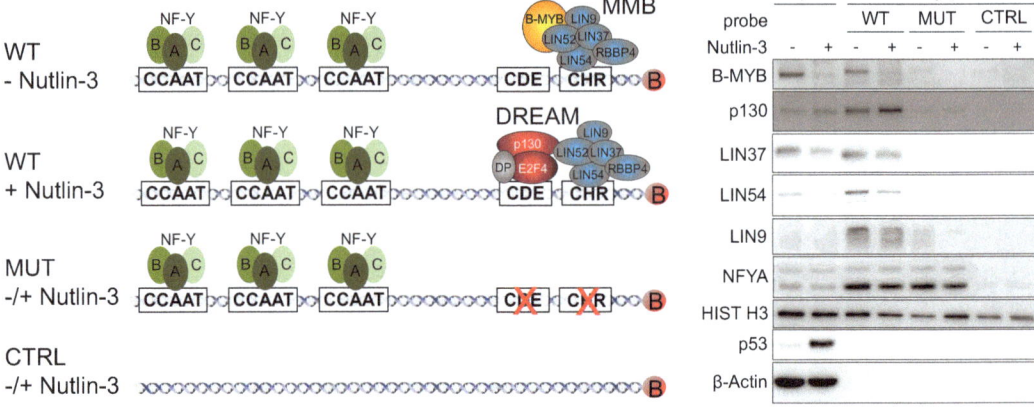

Fig. 1 Identification of proteins binding to CDE/CHR promoter elements of the *Cyclin B2* promoter with or without induction of the p53 signaling pathway. (**a**) Analysis of the *Cyclin B2* promoter with the UCSC genome browser [12] reveals phylogenetic conservation of CCAAT-boxes and of a CDE/CHR tandem element close to the transcription start site. (**b**) Induction of a p53 response by Nutlin-3 favors formation of the transcriptional repressor complex DREAM by activation of the CDK inhibitor p21. DREAM binds to CDE/CHR elements. In proliferating cells, the activating B-MYB-MuvB (MMB) complex also binds to these elements. Thus, DNA probes covering the basal *Cyclin B2* promoter (WT) allow for the purification of DREAM and MMB components. Binding of the DREAM-specific protein p130 is increased in Nutlin-3-treated samples, while of binding the MMB-specific protein B-MYB is decreased. DREAM or MMB components neither bind to *Cyclin B2* promoter probes with mutated CDE/CHR sites (MUT) nor to an irrelevant probe amplified from the pGL4.10 vector backbone (CTRL). Therefore, the CDE/CHR element can be identified as the central binding site for both complexes. In contrast, binding of the NF-Y subunit NF-YA is independent of CDE/CHR sites or p53

2.2 Nuclei Lysis Buffer

Nuclei Lysis Buffer: 20 mM HEPES, 1.5 mM MgCl$_2$, 0.2 mM EDTA, pH 7.9. To prepare 500 ml of Nuclei Lysis Buffer, weigh 2.38 g HEPES, 0.15 g MgCl$_2$·6H$_2$O, and 0.03 g EDTA and dissolve as in the previous step. Adjust pH with KOH. Store at 4 °C or at −20 °C.

2.3 Washing Buffer

To an appropriate volume of Nuclei Lysis Buffer, add NaCl to a final concentration of 150 mM (5 M stock solution in water) and 0.05% NP40 (stock solution 10% in water).

2.4 Elution Buffer

Dilute 5× Laemmli Buffer with Washing Buffer 1:5 to obtain 1× Elution Buffer.

5× Laemmli Buffer: 125 mM Tris–HCl pH 6.8, 25% Glycerol, 25% 2-mercaptoethanol, 10% SDS, 0.25% Bromophenol Blue. To prepare 16 ml of 5× Laemmli Buffer, add 0.04 g Bromophenol Blue and 1.6 g SDS to 6.0 ml ultrapure water, 2.0 ml 1 M Tris–HCl pH 6.8, 4.0 ml glycerol, and 4.0 ml 2-mercaptoethanol.

3 Methods

3.1 Preparation of Biotinylated Double-Stranded DNA Probes

1. Clone the promoter fragment to be analyzed into a vector of your choice (*see* **Note 1**). Mutate the putative protein binding site (e.g., by following the QuikChange mutagenesis protocol from Agilent).

2. Create PCR primers that cover the region that includes the promoter elements to be analyzed. Order unmodified forward PCR primers and add a 5′-biotin-modified nucleotide to the reverse primers. As a negative control, amplify a fragment of the vector backbone that covers approximately the same length (*see* **Note 2**).

3. Amplify probes by standard PCR. Prepare 800 μl reaction mixture per probe. Transfer the mastermix to PCR tubes so that each contains 100 μl of the reaction mixture. Run the PCR (*see* **Note 3**).

4. Prepare a large agarose gel. Depending on the length of the amplicon, use 1–2% agarose. Apply a preparative comb with an outer well reserved for the DNA size marker (*see* **Note 4**).

5. After the PCR is finished, pool all reactions from one probe. Add SYBR Green and DNA loading dye. If the amplicon is smaller than 200 bp, use a loading dye without Bromophenol Blue for a better visibility of the PCR product. Pipet the complete reaction mixture in the preparative well. Add DNA marker to the outer well.

6. Run the gel until the amplified DNA probe can be identified as a single band and is completely separated from any primer dimers. Cut out the band using a scalpel or a razor blade. Collect gel fragments in a 15 ml conical tube and determine the weight of the gel slices.

7. Purify the DNA probe using the Qiagen QIAEX II Gel Extraction Kit following the manufacturer's advice unless indicated otherwise. Add 3 ml buffer QX1 per 1 g of the gel and 30 μl QIAEXII suspension. Incubate for 60 min at 50 °C in a shaker (gel slices need to be completely dissolved). Before spinning down the beads, cool down tubes to room temperature to avoid deformation of the tubes during centrifugation. Spin

down beads for 10 min at 3000 × g at room temperature. Discard supernatant and resuspend pellets in 500 µl QX1. Transfer bead suspension to a 2 ml Eppendorf tube. Continue following the manufacturer's protocol. Elute DNA in 40 µl DNase-free ultrapure water for 5 min at 50 °C. Repeat elution step with additional 40 µl ultrapure water. Combine eluates.

8. Determine DNA concentration. The expected concentration is between 100 and 400 ng/µl.

3.2 DNA Affinity Purification

1. Cultivate sufficient cells to obtain at least 1 mg of nuclear extract for each probe (*see* **Note 5**).

2. Wash cells twice with cold PBS (kept at 4 °C). Use a cell scraper to collect cells in a 50 ml conical tube. Spin down cells at a maximum of 1000 × g for 5 min at 4 °C. Determine volume of the cell pellet.

3. Prepare 50 volumes of Nuclei Extraction Buffer per 1 volume of cells by adding protease inhibitors, phosphatase inhibitors, and DTT (1 M stock solution, 1:1000). Chill on ice (*see* **Note 6**).

4. Thoroughly resuspend cells in cold Nuclei Extraction Buffer. Incubate 20 min on ice. Vortex at full speed for 15 s.

5. Spin down nuclei at 2000 × g for 10 min at 4 °C. Discard supernatant to remove the cytosolic fraction.

6. Prepare Nuclei Lysis Buffer by adding protease inhibitors, phosphatase inhibitors and DTT (1 M stock solution, 1:1000). Chill on ice (*see* **Note 7**).

7. Thoroughly resuspend nuclei in cold Nuclei Lysis Buffer. Determine volume of resuspended nuclei (*see* **Note 7**).

8. Add NaCl to a final concentration of 450 mM (stock solution 5 M 1:10.1) to disrupt DNA–protein interactions.

9. Incubate for 15 min on ice.

10. Disrupt nuclei by passing through a syringe with a 23 gauge needle. Draw up the cell suspension into the syringe and eject with a single rapid stroke. Repeat five times. Avoid foam formation. If the extract is too viscous, start with a larger needle. Alternatively, the cells can be disrupted by ultrasonic lysis, for example, in an ultrasonic cup horn. Make sure the lysate stays cool during the sonication process.

11. Transfer the disrupted nuclei to 2 ml microcentrifuge tubes. Spin down at maximum speed ($\geq$10,000 × g) for 10 min at 4 °C to remove debris. Transfer the supernatants to new 2 ml microcentrifuge tubes. Repeat the centrifugation step. Collect lysates in 15 ml conical tubes if the volume is >500 µl,

otherwise continue using 2 ml microcentrifuge tubes. Determine volume of the nuclear extracts.

12. Add two volumes of Nuclei Lysis Buffer to one volume of lysate to adjust concentration of NaCl to 150 mM.

13. Determine the protein concentration of the cell lysate by Bradford assay. The concentration should be between 1 and 2 µg/µl. If the concentration is higher than 2 µg/µl, dilute with Nuclei Lysis Buffer w/150 mM NaCl. If the concentration is lower than 0.75 µg/µl, prepare new lysate using more cells or less lysis buffer.

14. Add sonicated salmon sperm DNA to a final concentration of 100 µg/ml to block unspecific DNA–protein interactions. Incubate on a rotator or tumbler for 15 min at room temperature.

15. Spin down at maximum speed for 1 min to remove precipitates.

16. Transfer 1 mg of nuclear extract for each DNA probe to 2 ml microcentrifuge tubes. If protein binding to DNA probes from different protein samples should be compared, use constant amounts of protein and adjust protein concentrations of all samples with Nuclei Lysis Buffer w/150 mM NaCl. Save at least 100 µg nuclear extract as input controls and store on ice.

17. Add 1 µg of DNA probe to 1 mg of nuclear extract (*see* **Note 8**).

18. Incubate on a rotator or tumbler for 30 min at room temperature.

19. If precipitates become visible, spin down again at maximum speed ($\geq$10,000 × g) for 1 min at 4 °C and transfer supernatants to new tubes.

20. Add streptavidin-coated magnetic beads. We recommend using Streptavidin MicroBeads from Miltenyi Biotec. Incubate 5 min at room temperature (*see* **Note 9**). Prepare 2 ml Washing Buffer for each sample by supplementing Nuclei Lysis Buffer with NaCl (final concentration 150 mM) and NP40 (final concentration 0.05%).

21. If working with the Miltenyi system, use the µMACS Separator in combination with µColumns to collect biotinylated probes following the manufacturer's instructions.

22. Wash columns 8× with 200 µl Washing Buffer. Pre-heat 150 µl Elution Buffer to 95 °C for each sample (*see* **Note 10**).

23. Add 20 µl pre-heated Elution Buffer to the columns. Use a new pipette tip for each sample to ensure constant volumes. Incubate 5 min.

24. Elute proteins by adding 100 μl pre-heated Elution Buffer to the columns and collect eluates in 1.5 ml microcentrifuge tubes. Use a new pipette tip for each sample.

25. Add the appropriate volume of 5× Laemmli Buffer to the input samples. Incubate inputs and eluates at 95 °C for 5 min. Store at −80 °C.

3.3 Protein Analysis Analyze samples by SDS-PAGE/Western blot or mass spectrometry. For Western blot analyses, run 19 μl of eluates and 15 μg of input samples on SDS-PAGE. Thus, one purification is sufficient for 5 blots.

4 Notes

1. To be able to test the effect of identified proteins and binding sites on promoter activity in luciferase reporter assays, use a promoterless firefly luciferase reporter vector to clone the fragment to be analyzed (e.g., pGL4.10, Promega). The cloned fragment should cover at least the basal promoter. If aiming to discover unknown regulating elements in a specific promoter, apply the conservation tracks in the UCSC Genome Browser [12] to identify evolutionary conserved promoter regions. Such conserved elements are generally essential for proper transcriptional regulation (Fig. 1). To analyze protein binding to specific DNA elements, create promoter mutants with non-functional elements (e.g., following the QuikChange protocol). Test the impact of the mutations in luciferase promoter reporter assays [4]. Furthermore, regulation of promoters may be analyzed by knockdown or overexpression of specific transcription factors.

2. The optimal length of DNA probes is 200 to 500 bp. However, we have obtained good results with probes between 0.1 and 1 kb in length. To reduce costs when preparing multiple probes, amplify the complete inserts with PCR primers binding to regions in the plasmid directly upstream and downstream of the promoter inserts. The putative protein binding site to be investigated does not have to be located at the center of the probe; however, there should be a spacer of at least 20 bp between the element and the end of the probe.

3. For amplification of the probes, standard Taq polymerase is generally sufficient and most cost efficient. Add 500 ng template DNA and 3.2 μl of forward and reverse primers (100 μM each). Run PCR with 37 cycles. As promoter regions tend to have a high GC content, adding 5% DMSO to the reaction mixture often helps to increase the yield of the PCR product.

4. If a large preparative comb is not available, adhesive tape may be used to convert a multiple-well comb into a preparative comb.

5. The number of cells needed is highly dependent on cell size and the culture conditions as well as on the number of the DNA probes. Before preparing cells for a DNA affinity purification assay, consider performing a preliminary experiment to determine the yield of nuclear extract from a specific number of cells. Generally, 150 mm cell culture dishes are a good choice for cultivation of the cells. The yield of nuclear proteins obtained from average sized cells collected from a 150 mm plate at 90% confluency is usually between 1 and 2 mg. When DNA–protein interactions under different cellular conditions are compared (e.g. proliferating cells versus serum-starved cells), make sure to prepare sufficient cells for each condition.

6. The DNA affinity purification can also be performed with whole cell extract. However, background protein signals may be higher. If the amount of cells is limited or if cells are frozen prior to lysis, the nuclear extraction step can be omitted and cells can be directly resuspended in Nuclei Lysis Buffer.

    Take into account to prepare sufficient Nuclei Lysis Buffer for each purification: To lyse the nuclei in High Salt Buffer (**step 7**), to dilute nuclear extracts to physiological salt conditions (**step 12**), to adjust protein concentration and volumes of different samples to allow comparability (**step 16**) and to prepare Washing Buffer (8 × 200 µl, **step 22**). Generally, 4 ml Nuclei Lysis Buffer for a single purification reaction are sufficient. However, if nuclear extracts are prepared from large numbers of cells, it may be necessary to increase the volume.

7. The volume of Nuclei Lysis Buffer needed depends on the amount and the size of nuclei and the number of samples. For each cell line, the optimal relation between the volume of collected nuclei and Nuclei Lysis Buffer has to be determined. Usually, resuspending nuclei in 3 volumes of Nuclei Lysis Buffer is a good starting point.

8. The reactions can easily be scaled up. If the purified proteins should be analyzed by mass spectrometry, we suggest preparing reaction samples of 5 mg nuclear extract and 5 µg DNA probe. For Western blot analyses, 1 mg extract and 1 µg probe is usually sufficient.

9. We generally utilize the µMACS™ Streptavidin Kit from Miltenyi Biotec as handling is very easy and background signals are very low. 30 µl Streptavidin MicroBeads suspension is generally sufficient to isolate proteins bound to 1 µg biotinylated DNA probe. Coated magnetic beads from other suppliers (e.g.,

streptavidin-coupled Dynabeads® from Thermo Fisher Scientific) should work as well. We do not recommend using Sepharose or agarose beads.

10. If beads from other suppliers are utilized and washing is not performed in columns but in microcentrifuge tubes, it may be necessary to adjust the volumes of Washing Buffer and the number of washing steps. Since some proteins bind to DNA in a non-sequence-specific manner, background binding may be reduced by increasing the concentration of NaCl or NP40 in the Washing Buffer.

References

1. Zwicker J, Lucibello FC, Wolfraim LA, Gross C, Truss M, Engeland K, Müller R (1995) Cell cycle regulation of the cyclin A, cdc25C and cdc2 genes is based on a common mechanism of transcriptional repression. EMBO J 14:4514–4522

2. Müller GA, Wintsche A, Stangner K, Prohaska SJ, Stadler PF, Engeland K (2014) The CHR site: definition and genome-wide identification of a cell cycle transcriptional element. Nucleic Acids Res 42:10331–10350

3. Müller GA, Engeland K (2010) The central role of CDE/CHR promoter elements in the regulation of cell cycle-dependent gene transcription. FEBS J 277:877–893

4. Müller GA, Quaas M, Schümann M, Krause E, Padi M, Fischer M, Litovchick L, DeCaprio JA, Engeland K (2012) The CHR promoter element controls cell cycle-dependent gene transcription and binds the DREAM and MMB complexes. Nucleic Acids Res 40:1561–1578

5. Mages CF, Wintsche A, Bernhart SH, Müller GA (2017) The DREAM complex through its subunit Lin37 cooperates with Rb to initiate quiescence. Elife 6:e26876

6. Litovchick L, Sadasivam S, Florens L, Zhu X, Swanson SK, Velmurugan S, Chen R, Washburn MP, Liu XS, DeCaprio JA (2007) Evolutionarily conserved multisubunit RBL2/p130 and E2F4 protein complex represses human cell cycle-dependent genes in quiescence. Mol Cell 26:539–551

7. Fischer M, Müller GA (2017) Cell cycle transcription control: DREAM/MuvB and RB-E2F complexes. Crit Rev Biochem Mol Biol 52:638–662

8. Müller GA, Stangner K, Schmitt T, Wintsche A, Engeland K (2017) Timing of transcription during the cell cycle: protein complexes binding to E2F, E2F/CLE, CDE/CHR, or CHR promoter elements define early and late cell cycle gene expression. Oncotarget 8:97736–97748

9. Mannefeld M, Klassen E, Gaubatz S (2009) B-MYB is required for recovery from the DNA damage-induced G2 checkpoint in p53 mutant cells. Cancer Res 69:4073–4080

10. Quaas M, Müller GA, Engeland K (2012) p53 can repress transcription of cell cycle genes through a p21(WAF1/CIP1)-dependent switch from MMB to DREAM protein complex binding at CHR promoter elements. Cell Cycle 11:4661–4672

11. Engeland K (2018) Cell cycle arrest through indirect transcriptional repression by p53: I have a DREAM. Cell Death Differ 25:114–132

12. Kent WJ, Sugnet CW, Furey TS, Roskin KM, Pringle TH, Zahler AM, Haussler D (2002) The human genome browser at UCSC. Genome Res 12:996–1006

Chapter 7

A Novel Strategy to Track Lysine-48 Ubiquitination by Fluorescence Resonance Energy Transfer

Kenneth Wu and Zhen-Qiang Pan

Abstract

Posttranslational modification of protein by lysine-48 (K48) linked ubiquitin (Ub) chains is the major cellular mechanism for selective protein degradation that critically impacts biological processes such as cell cycle checkpoints. In this chapter, we describe an in vitro biochemical approach to detect a K48-linked di-Ub chain by fluorescence resonance energy transfer (FRET). To this end, we detail methods for the preparation of the relevant enzymes and substrates, as well as for the execution of the reaction with high efficiency. Tracking K48 polyubiquitination using this sensitive and highly reproducible format provides an opportunity for high-throughput screening that leads to identification of small molecule modulators capable of changing ubiquitination for improving human health.

Key words K48 ubiquitination, Cullin-RING E3 ubiquitin ligase, E2 Cdc34, Fluorescently labeled ubiquitin, FRET

1 Introduction

Timely removal of cell cycle regulators by ubiquitin (Ub)-dependent proteolysis is fundamental to the functioning of cell division cycle [1]. Central to selective protein turnover by the 26S proteasome is the formation of homotypic Lysine-48 (K48) linked Ub chains that tag substrate proteins for degradation [2]. The underlying biochemical process, termed ubiquitination, is driven by three enzymes: an E1 activating enzyme, an E2 conjugating enzyme, and an E3 ligase that recognizes a substrate and hence governs the specificity of the reaction [2]. One well-characterized K48-polyubiquitination system involves E3 Cullin-RING Ub ligase 1 (CRL1; also named as SCF [Skp1-Cullin1-F-box-ROC1/Rbx1] complex) and E2 Cdc34 [3, 4]. CRL1 substrates include many key cell cycle modulators, such as p27 and cyclin E [5].

To track CRL1-mediated K48 ubiquitination by fluorescence, we developed a "fluorescence (Förster) resonance energy transfer (FRET) K48 di-Ub assay" [6]. The assay is composed of five

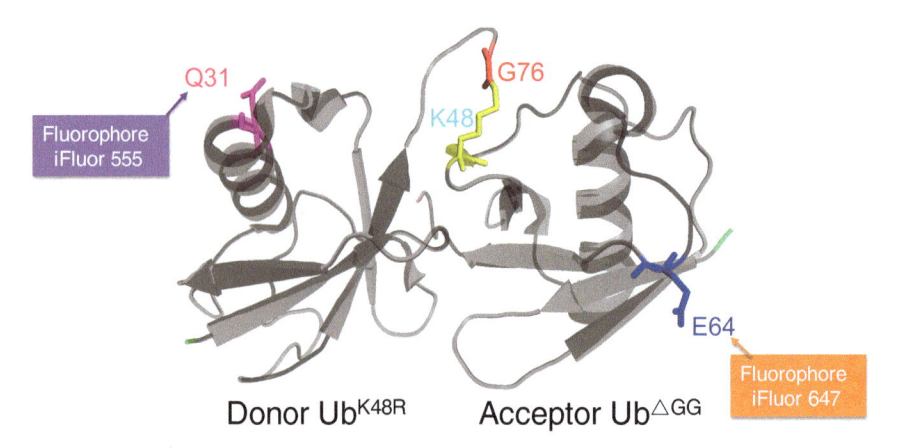

Fig. 1 Conjugation of fluorescent dye to a pair of residues on the donor and acceptor Ub. A K48 di-Ub chain is shown based on previous crystallographic studies [16] (PDB 1F9J). At position Q31 or E64 of the donor or acceptor Ub, respectively, each residue is substituted by cysteine and labeled with the indicated fluorescent dye. For iFluor 555, Exmax is 559 nm and Emmax is 569 nm. For iFluor 647, Exmax is 654 nm and Emmax is 674 nm. The calculated distance between Ub C31-linked iFluor 555 and Ub C64-linked iFluor 647 in the K48 di-Ub chain is 38 Å. This distance is within the range of the Förster radius (Ro) for iFluor 555/iFluor 647, which is about 51 Å (at this distance the resonance transfer efficiency is 50%)

components: (1) E1, (2) E2 Cdc34 specific for K48 linkage, (3) E3 sub-complex ROC1-Cullin 1 (CUL1) (411–776) (called ROC1-CUL1 CTD) required for E2 activation, (4) donor Ub (carrying K48R substitution and fluorescent dye iFluor 555 at position Q31C; Fig. 1), and (5) receptor Ub (bearing G75G76 deletion and fluorescent dye iFluor 647 at position E64C; Fig. 1). The reaction has been modified from previously reported versions [7], allowing for only one nucleophilic attack to produce a single Ub–Ub isopeptide bond (Fig. 2) exclusively carrying a K48-linkage [8]. In this ubiquitination system, ROC1 binds to the Cdc34 catalytic core domain [9], whereas the CUL1 CTD provides a basic pocket that interacts with Cdc34's acidic C-terminus, thereby facilitating E2 recruitment [10].

The "FRET K48 di-Ub assay" was effectively used for a high-throughput screen (HTS) campaign to discover small molecule chemical probes that target E3 CRL1 [6]. The restriction to a single Ub–Ub linkage eliminates complexity associated with polyubiquitin chains, ensuring a high degree of reproducibility that is critical to HTS success. Secondly, FRET is produced by two Ub-linked fluorophores that become juxtaposed upon Ub–Ub conjugation (Fig. 2). Each fluorophore is uniquely engineered to either donor or receptor Ub at specific site (Fig. 1). Note that optimal positioning of fluorophore in Ub is critical, as N-terminally labeled Ub, commonly used in previous publications [11, 12] is inactive in our system. Commercial fluorescent Ub proteins do not specify the location of the fluorescent dye in Ub and it is unclear if such reagents can be generally optimal for HTS

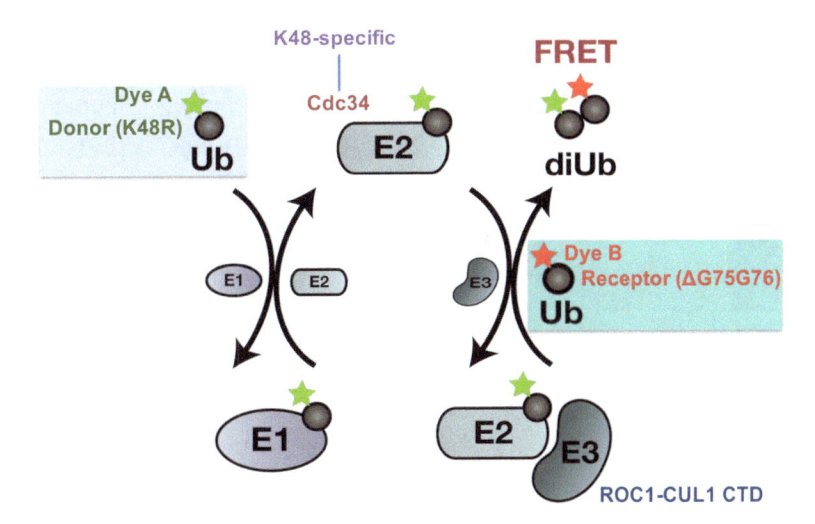

Fig. 2 FRET K48 di-Ub assay Scheme. E1 and UbQ31C/K48R-iFluo 555 form a thiolester complex. Transthiolation then follows to yield Cdc34~UbQ31C/K48R-iFluo 555. In the presence of E3 sub-complex ROC1-CUL1 CTD, the K48 residue of UbE64C/ΔGG-iFluo 647 attacks Cdc34~UbQ31C/K48R-iFluo 555. Consequently, the resulting di-Ub brings iFluo 555 and iFluo 647 into proximity that generates FRET upon donor fluorophore excitation

FRET assays. On the other hand, this FRET strategy should be applicable to other HTS campaigns that depend on K48 ubiquitination, but require different E3/E2 enzyme combinations.

The method detailed here provides a FRET-based strategy for accurately quantifying K48 polyubiquitination, and can be used as a platform to discover and characterize small molecule agents capable of perturbing ubiquitination.

2 Materials

2.1 Preparation of Ub, E1, and E2 Cdc34 (See Note 1)

1. pHisTEVyUb-Q31C/K48R (*see* **Note 2**), pHisTEVyUb-E64C/ΔGG (*see* **Note 3**), pET3a-Uba1 (E1)-His10 (*see* **Note 4**), or pET28a-His6-TEV Cdc34 C191S/C223S (*see* **Note 5**) expression plasmid.

2. *Escherichia coli* Rosetta 2(DE3)pLysS cells, stored at −80 °C (EMD Millipore).

3. Luria–Bertani (LB) Broth Base, diluted to 15.5 g/L, autoclaved, and stored at 4 °C (Sigma-Aldrich).

4. Carbenicillin, stock is 100 mg/mL in H_2O, and stored at −20 °C (Sigma-Aldrich).

5. LB Agar Plates + carbenicillin (100 μg/mL) (Sigma-Aldrich). Isopropyl β-D-1-thiogalactopyranoside (IPTG), stock is 1 M in H_2O (Sigma-Aldrich).

6. Lysis buffer: 20 mM Tris–HCl, pH 8.0, 1% Triton X-100, 0.5 M NaCl, 2 mM phenylmethylsulfonyl fluoride, 0.4 μg/mL antipain, and 0.2 μg/mL leupeptin).

7. Buffer A: 25 mM Tris–HCl (pH 7.5), 0.01% Nonidet P-40 (NP-40), 10% glycerol, 1 mM dithiothreitol (DTT), 0.1 mM PMSF, and 0.2 μg/mL of antipain and leupeptin).

8. Imidazole wash buffer: Buffer A (2.1.7) + 50 mM NaCl + 5 mM imidazole.

9. Imidazole elution buffer: Buffer A (2.1.7) + 50 mM NaCl + 250 mM imidazole.

10. Ni-nitrilotriacetic acid-agarose (Ni-NTA) beads (Qiagen). Dialysis buffer: 25 mM Tris–HCl, pH 7.4, 10% glycerol, 50 mM NaCl, 0.01% NP-40, and 1 mM DTT.

11. SnakeSkin Pleated Dialysis Tubing, 3500 MWCO (Pierce).

12. Mono Q HR 10/10 column (Amersham Biosciences).

13. Amicon Ultra-15 Centrifugation Filters, 10 K (Millipore).

14. Sephadex-75 gel filtration column (Amersham Biosciences).

15. AKTA FPLC (GE Healthcare).

2.2 Preparation of E3 ROC1-CUL1 CTD Complex

1. pETDuet-1-MCS-I-CUL1 (L421E, V451E, V452K, Y455K)/MCS-II-ROC1 expression plasmid (*see* **Note 6**).

2. BL21-CodonPlus(DE3)-RIL cells, stored at −80 °C (Agilent). Wash buffer (50 mM Na_2HPO_4, 300 mM NaCl, 10 mM imidazole, pH 8.0).

3. EDTA-free protease inhibitor tablet (Roche).

4. EmulsiFlex-C5 homogenizer (Avestin, Ottawa, Ontario).

5. Filter (0.45μm, Millipore, Mississauga, ON, Canada).

6. HisTrap FF column (GE Healthcare).

7. Elution buffer (50 mM Na_2HPO_4, 300 mM NaCl, and 250 mM imidazole, pH 8.0).

2.3 Preparation of Fluorescently Labeled Ub

1. Anaerobic glove box (Captair Pyramid, Erlab) under nitrogen gas.

2. TCEP (Sigma).

3. Phosphate-buffered saline (PBS).

4. iFluor 555 maleimide and iFluor 647 maleimide (AAT Bioquest).

5. DMF (Sigma).

2.4 FRET K48 di-Ub Assay

1. Costar 384 well storage plate #3658 (Corning).

2. Ub E1 (Subheading 3.1), stored at −80 °C.

3. E2 Cdc34 (Subheading 3.1), stored at −80 °C.

4. ROC1-CUL1 CTD (Subheading 3.2), stored at −80 °C.
5. Ub C31-iFluor 555 (Subheading 3.3) and Ub C64-iFluor 647 (Subheading 3.3), stored at 4 °C.
6. Centrifuge with adaptor for microtiter plate (Eppendorf).
7. Synergy-H1 reader (BioTek).

2.5 Running the Gel

1. NOVEX 4–20% Bis-Tris Protein Gel (12–15 well, 1.0 mm).
2. 10× Tris/Glycine/SDS Running Buffer, diluted to 1× with distilled H_2O (Bio-Rad).
3. Protein marker (Bio-Rad).
4. Power supply (Bio-Rad).
5. 4× SDS loading buffer: 40% glycerol, 240 mM Tris–HCl (pH 6.8), 8% SDS, and 0.04% bromophenol blue dissolved in H_2O. Coomassie Brilliant Blue solution: Coomassie R-250 (Bio-Rad), methanol (30% v/v), glacial acetic acid (10% v/v).
6. Destaining solution: methanol (10% v/v), glacial acetic acid (10% v/v).

2.6 Fluorescence Imaging

1. Typhoon FLA 9500 laser scanner (GE).

---

3 Methods

3.1 Preparation of Ub, E1, or E2 Cdc34

1. Transform pHisTEVyUb-Q31C/K48R, pHisTEVyUb-E64C/ΔGG, pET3a-Uba1 (E1)-His10, or pET28a-His6-TEV Cdc34 C191S/C223S plasmid into *Escherichia coli* Rosetta 2(DE3)pLysS cells. Incubate cells on ice for 30 min, heat-shock at 37 °C for 30 s, and return to ice for 2 min. Plate on LB agar plates with carbenicillin.
2. Pick a single colony and inoculate in a starter culture overnight in LB + 100 μg/mL carbenicillin (*see* **Note 7**).
3. Add the starter culture to the growth culture (LB with 0.4% glucose in the presence of carbenicillin) in a 1:100 ratio of dilution. Grow at 37 °C until the optical density measured at 600 reaches 0.5 (~3–3.5 h).
4. Induce culture with IPTG at a final concentration of 1 mM overnight at 25 °C.
5. Pellet bacterial cells by centrifugation at 5000 × *g* for 15 min at 4 °C.
6. Resuspend cells in 1/25 the culture volume of lysis buffer (Subheading 2.1, **item 6**).
7. Sonicate the lysate (4× 20 s) and centrifuge it at 10,000 × *g* for 30 min.

8. Mix soluble extracts with Ni-nitrilotriacetic acid-agarose beads (Ni-NTA) beads, at a ratio of 0.3 mL beads per 1 mL extracts, and rotate the mixture for 2 h at 4 °C.

9. Pack beads into a Flex-Column and wash with buffer A (Subheading 2.1, **item 7**) + 50 mM NaCl, in 10× the column volume, and imidazole wash buffer (Subheading 2.1, **item 8**), in 3× the column volume.

10. Elute protein with imidazole elution buffer (Subheading 2.1, **item 9**), in 10× the column volume, collecting 1 mL fractions.

11. Determine peak fractions by subjecting aliquots of the fractions (1μL) to SDS-PAGE (Subheading 2.5). Stain the gel for 10 min in Coomassie solution (Subheading 2.5, **item 5**), rinse with distilled water, and destain in destaining solution (Subheading 2.5, **item 6**).

12. Pool peak fractions, load into dialysis tubing prerinsed with distilled water, and properly seal the ends. Dialyze twice, 5 h to overnight each time, by placing the tubing in 10–20× the volume of dialysis buffer at 4 °C (Subheading 2.1, **item 10**).

13. No further steps are necessary for Ub-Q31C/K48R or Ub-E64C/ΔGG. The yield is approximately 83 or 30 mg of purified Ub-Q31C/K48R or Ub-E64C/ΔGG (from 1 L culture), respectively. They are stored at −80 °C.

14. To further purify E1, load Ni-NTA-purified, dialyzed material onto a Mono Q HR 10/10 column using the AKTA FPLC. Wash column with buffer A + 50 mM NaCl.

15. Elute the protein with a 20 column volume linear gradient of 50–500 mM NaCl in buffer A (Subheading 2.1, **item 7**). Analyze by SDS-PAGE as in **step 11** and pool the peak fractions.

16. Concentrate the peak fractions using centrifugal filters. The yield is approximately 4 mg of purified E1 from 1 L culture (*see* **Note 8**).

17. To purify Cdc34, follow **steps 14–16**. The expected yield for Cdc34 is approximately 18 mg from 1 L of culture (*see* **Note 8**).

3.2 Preparation of E3 ROC1-CUL1 CTD Complex

1. Transform pETDuet-1-MCS-I-CUL1 (L421E, V451E, V452K, Y455K)/MCS-II-ROC1 plasmid into BL21-Codon-Plus(DE3)-RIL cells. Incubate cells on ice for 30 min, heat-shock at 37 °C for 30 s, and return to ice for 2 min. Plate on LB agar plates with carbenicillin.

2. Pick a single colony and inoculate in a starter culture overnight in LB + 100 μg/mL carbenicillin + 34 mg/L chloramphenicol (*see* **Note 7**).

3. Add 10 mL of the starter bacterial colony to 4 L (1:400 dilution) of prewarmed LB media supplemented with 0.5 mM ZnCl$_2$, 100 mg/L ampicillin, and 34 mg/L chloramphenicol. Grow at 37 °C at 210 rpm. When the culture reaches an OD600 of 0.4, the temperature is reduced to 16 °C with continued shaking.

4. Once the OD600 reached 0.7, the culture is induced with 1 mM IPTG and the cells are grown overnight.

5. Pellet the cells by centrifugation at 6000 × *g* for 10 min at 4 °C. Resuspend cells in 25 mL wash buffer with an EDTA-free protease inhibitor tablet.

6. Lyse cells using an EmulsiFlex-C5 homogenizer. If the mixture is still too viscous, pass the lysate through a syringe with a 21G syringe several times.

7. Clear lysates by centrifugation (110,000 × *g* for 1 h at 4 °C).

8. Filter the supernatant.

9. Load the filtered extracts onto a 5 mL HisTrap FF column preequilibrated with wash buffer at a flow rate of 0.5 mL/min using an AKTA FPLC.

10. Wash the column at 3 mL/min (15 column volumes with wash buffer containing 30 mM imidazole, then 10 column volumes with 60 mM imidazole).

11. Elute the ROC1-CUL1 CTD complex with elution buffer (Subheading 2.2, **item 7**) at a flow rate of 2 mL/min. Fractions containing the ROC1-CUL1 CTD complex, as determined by staining analysis following protein separation by SDS-PAGE, as in Subheading 3.1, **step 11**, were pooled.

12. Dialyze the pooled fraction as in Subheading 3.1, **step 12**.

13. To further purify ROC1-CUL1 CTD, proceed to **steps 14–16** in Subheading 3.1.

14. The yield for the ROC1-CUL1 CTD complex is approximately 48 mg from 1 L of culture.

15. Protein aliquots are stored at −80 °C (*see* **Note 8**).

3.3 Preparation of Fluorescently Labeled Ub

1. **Steps 1–4** manipulations are carried out inside an anaerobic glove box. To prepare fluorescent Ub, purified Ub-Q31C/K48R and Ub-E64C/ΔGG, 150 mg each, were treated with 10 mM TCEP in degassed phosphate-buffered saline (PBS) for 15 min at room temperature.

2. Add Ni-NTA agarose for 1 h at 4 °C in sealed tubes.

3. Wash the beads with degassed PBS to remove the TCEP and any unbound protein.

4. Incubate the resulting beads with 15 mg of either iFluor 555 maleimide for UbQ31C/K48R, or iFluor 647 maleimide for UbE64C/ΔGG), dissolved in DMF, for 1 h at room temperature then overnight at 4 °C, in the dark.

5. The subsequent steps were performed in normal air. Following incubation, beads were first washed with PBS to remove unreacted dye.

6. Elute the labeled proteins with PBS plus 250 mM Imidazole.

7. Dialyze the eluted proteins at 4 °C in the dark, against buffer A (Subheading 2.1, **item 7**) plus 150 mM NaCl. The final yield is ~108 mg or ~129 mg for labeled Ub-Q31C/K48R or Ub-E64C/ΔGG, respectively (*see* **Note 9**).

3.4 FRET K48 di-Ub Assay

1. Thaw all reagents on a metal block on ice (*see* **Note 10**).

2. The reaction mixture (15μL) is assembled onto a 384-well microtiter plate (or in a test tube). Each well contained 33 mM Tris–HCl (pH 7.4), 1.7 mM MgCl$_2$, 0.33 mM DTT, BSA (0.07 mg/mL), Ub E1 (14 nM), E2 Cdc34 (124 nM), ROC1-CUL1 CTD (1μM), Ub C31-iFluor 555 (donor, 0.93μM), and Ub C64-iFluor 647 (receptor, 1.62μM) (*see* **Note 11**).

3. ATP (0.66 mM) is then added to the mix followed by a brief centrifugation to bring down the mixture.

4. The resulting plate is incubated at 30 °C in the Synergy-H1 reader and the fluorescence signal is monitored. Ubiquitination is quantified based on the ratio of acceptor–donor fluorescence (excitation 515 nm; donor emission 570 nm, acceptor emission 670 nm).

5. Figure 3 provides FRET detection of K48-ubiqutination. The intensity of fluorescence signals is shown across a range of wavelength as indicated. The area representing FRET signals is marked.

6. In parallel, the reaction is analyzed by gel electrophoresis. For this purpose, an aliquot of final reaction mixture (2μL) is subject to SDS-PAGE (Subheading 3.5), followed by imaging with Typhoon FLA 9500 laser scanner (Subheading 3.6).

7. Figure 4 provides a fluorescent image of K48 ubiquitination. The positions of the fluorescent Ub input as well as the "di-Ub" product are marked. Omission of reaction components abolished ubiquitination. CC0651, a recently discovered Cdc34 inhibitor [13], is added to the reaction at concentration of 2, 10 or 100μM, resulting in inhibition of the ubiquitination.

8. *See* **Notes 12** to **14** for further result analysis.

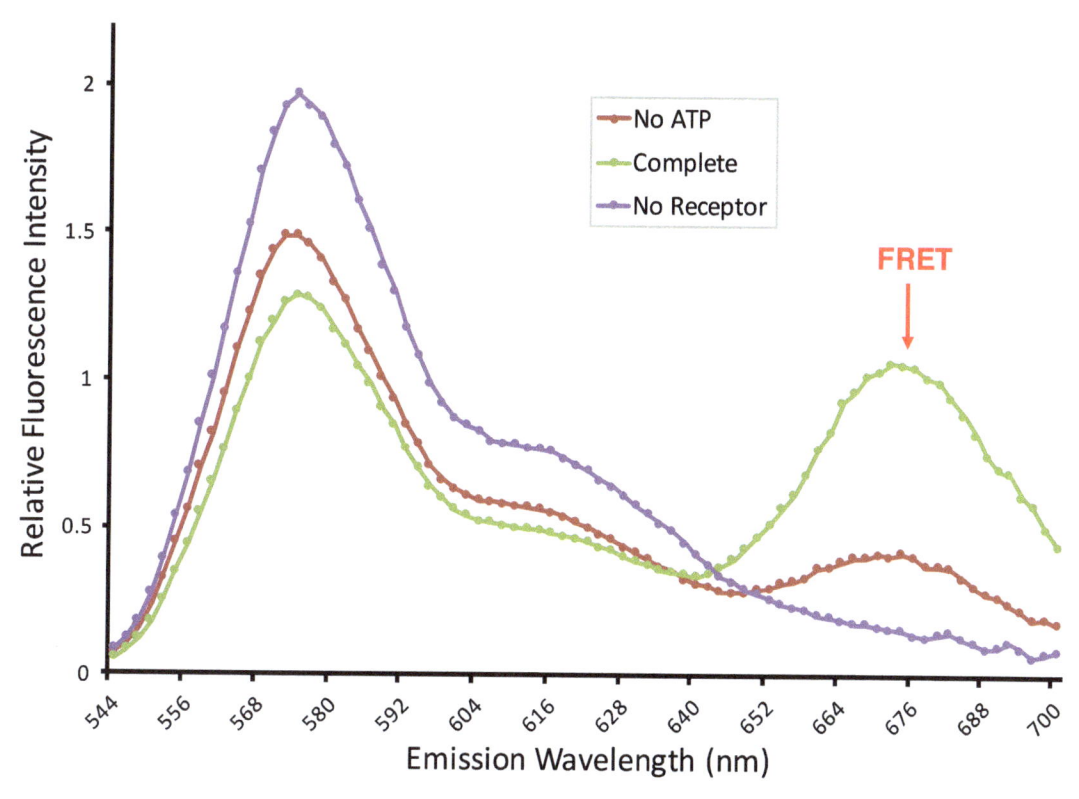

Fig. 3 Detect K48 ubiquitination by FRET. FRET K48 di-Ub assay was carried out as described in Subheading 3.4. The fluorescence signals ranging from 545 to 700 nm were shown. Only the complete reaction yielded FRET

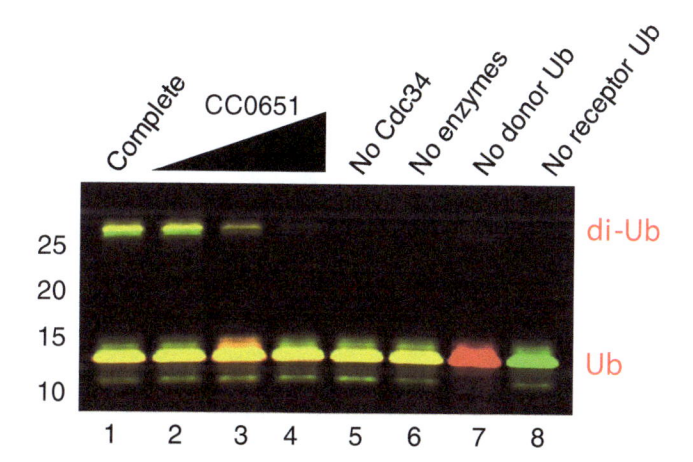

Fig. 4 Detect K48 ubiquitination by fluorescence imaging. Aliquots of the reaction products of the FRET K48 di-Ub assay was analyzed by SDS-PAGE (Subheading 3.5) and imaging (Subheading 3.6). Shown is an image of fluorescent reaction products detected by a Typhoon 9500 scanner. Only the complete reaction where all components of the assay were present supported di-Ub formation. In addition, the di-Ub ubiquitination was sensitive to the Cdc34 inhibitor CC0651 [13]

3.5 Running the Gel

1. Remove the gel from the wrapper, remove the comb, and flush the wells with water before placing in the gel apparatus.
2. Pour running buffer into the chamber and ensure that it covers the wells of the gel.
3. Boil the samples in SDS loading buffer (1× final concentration) for 5 min at 95 °C and centrifuge briefly afterward to collect the sample.
4. Load 2–5 μL of protein standard, along with your samples, using gel-loading tips.
5. Connect the power supply and run at 150 V until the loading dye approaches the bottom of the gel (or until the 10 kDa marker approaches the bottom).

3.6 Fluorescence Imaging of the Gel

1. When the gel is finished running, remove it from the cassette, cut off the wells and bottom of the gel, and place it into PBS in a tray.
2. Place the gel on a prewetted area in the glass screen within the Typhoon FLA 9500 laser scanner.
3. Scan the gel.

4 Notes

1. The E3 ROC1-CUL1 CTD complex and fluorescently labeled Ub-Q31C/K48R as well as Ub-E64C/ΔGG are not commercially available. While E1 activating enzyme and E2 enzyme Cdc34 can be purchased from Boston Biochem, we provide the protocols for the preparation of E1 and Cdc34 for cost-saving.
2. pHisTEVyUb-Q31C/K48R leads to bacterial expression of a variant form of Ub bearing Q31C and K48R substitutions. These changes allow the placement of a fluorophore at amino acid position C31 through maleimide linkage (Fig. 1). The resulting fluorescent Ub can only act as a donor in Cdc34-catalyzed ubiquitination due to the lack of K48 (Fig. 2). Additional information can be found in ref. 6.
3. pHisTEVyUb-E64C/ΔGG leads to bacterial expression of a variant form of Ub bearing E64C substitution and G75G76 deletion. These changes allow the placement of a fluorophore at amino acid position C64 through maleimide linkage (Fig. 2). The resulting fluorescent Ub can only act as a receptor due to the lack of C-terminal G75G76 motif (Fig. 2). Additional information can be found in ref. 6.
4. pET3a-Uba1 (E1)-His10 is described in Addgene (Plasmid #63571; https://www.addgene.org/63571/).

5. pET28a-His6-TEV Cdc34 C191S/C223S leads to bacterial expression of a variant form of Cdc34 bearing C191S and C223S substitutions [9]. This E2 variant is indistinguishable from the wild type in reconstituted ubiquitination assays.

6. pETDuet-1-MCS-I-CUL1 (L421E, V451E, V452K, Y455K)/MCS-II-ROC1 leads to bacterial coexpression of ROC1 and the C-terminal fragment of CUL1 amino acids 411–776. Substitutions L421E, V451E, V452K, Y455K are introduced to improve CUL1 protein solubility [14] without diminishing its activity in reconstituted ubiquitination assays.

7. Glycerol stocks can be made and stored long term at −80 °C.

8. Enzymes are stored at −80 °C. Multiple freeze–thaw cycles can decrease the activity of the enzyme, so it is best stored in small aliquots. All other enzymes should also be stored at −80 °C.

9. For long-term storage, fluorescently labeled Ubs are kept at −80 °C. For everyday use, an aliquot can be stored at 4 °C for a period of 6 months without significant loss of activity in ubiquitination.

10. It is important to only thaw enzymes immediately before use, and put them away immediately afterward.

11. The concentrations of all five protein reagents have been determined based on Km measurement.

12. As shown, di-Ub product of expected size formed only in the complete reaction (Fig. 4). Likewise, a Robust FRET signal is detected only when all five proteins were present (Fig. 3). Thus, the observed FRET truly reflects formation of the K48-linked di-Ub chain as a result of enzymatic synthesis. Note the optional positioning of fluorophore in Ub is critical, as N-terminally labeled Ub, commonly used in previous studies [11, 12], is inactive in our assay. Commercial fluorescent Ub proteins do not specify the locations of the fluorescent dye in Ub and it is unclear if such reagents can be generally optimal for this Ub K48-specific FRET assay.

13. Z' factor measures statistical effect size based on the means and standard deviations of both the positive and negative control samples. A Z' factor value in the range of 0.5–1 is indicative of an excellent assay for high-throughput screening [15]. As detailed in Ref. 6, the Z' factor of the FRET K48 di-Ub assay is 0.86, suggesting that this reporter system provides a sufficient screening window.

14. The reactions can be stored at −20 °C at this point, and can be run at a later time.

Acknowledgments

We thank R. Chong, Q. Yu, C. Lee, and other members of the Pan lab for their assistance with protocols. This work was supported by Public Health Service grants CA251425-01 and GM122751-01 to Z-Q P.

References

1. Koepp DM, Harper JW, Elledge SJJW (1999) How the cyclin became a cyclin: regulated proteolysis in the cell cycle. Cell 97(4):431–434
2. Hershko A, Ciechanover A (1998) The ubiquitin system. Annu Rev Biochem 67:425–479
3. Petroski MD, Deshaies RJ (2005) Function and regulation of cullin-RING ubiquitin ligases. Nat Rev Mol Cell Biol 6:9–20
4. Sarikas A, Hartmann T, Pan ZQ (2011) The cullin protein family. Genome Biol 12(4):220
5. Skaar JR, D'Angiolella V, Pagan JK, Pagano M (2009) SnapShot: F box proteins II. Cell 137 (7):1358–1358 e1351
6. Wu K, Chong RA, Yu Q et al (2016) Suramin inhibits cullin-RING E3 ubiquitin ligases. Proc Natl Acad Sci U S A 113(14):E2011–E2018
7. Wu K, Kovacev J, Pan Z-Q (2010) Priming and extending: a UbcH5/Cdc34 E2 handoff mechanism for polyubiquitination on a SCF substrate. Mol Cell 37:784–796
8. Chong RA, Wu K, Spratt DE et al (2014) Pivotal role for the ubiquitin Y59-E51 loop in lysine 48 polyubiquitination. Proc Natl Acad Sci U S A 111(23):8434–8439
9. Spratt DE, Wu K, Kovacev J et al (2012) Selective recruitment of an E2~ubiquitin complex by an E3 ubiquitin ligase. J Biol Chem 287 (21):17374–17385
10. Kleiger G, Saha A, Lewis S et al (2009) Rapid E2-E3 assembly and disassembly enable processive ubiquitylation of cullin-RING ubiquitin ligase substrates. Cell 139(5):957–968
11. Madiraju C, Welsh K, Cuddy MP et al (2012) TR-FRET-based high-throughput screening assay for identification of UBC13 inhibitors. J Biomol Screen 17(2):163–176
12. Carlson CB, Horton RA, Vogel KW (2009) A toolbox approach to high-throughput TR-FRET-based SUMOylation and DeSUMOylation assays. Assay Drug Dev Technol 7:348–355
13. Huang H, Ceccarelli DF, Orlicky S et al (2014) E2 enzyme inhibition by stabilization of a low-affinity interface with ubiquitin. Nat Chem Biol 10(2):156–163
14. Duda DM, Borg LA, Scott DC et al (2008) Structural insights into NEDD8 activation of Cullin-RING ligases: conformational control of conjugation. Cell 134(6):995–1006
15. Zhang J-H, Chung TDY, Oldenburg KR (1999) A simple statistical parameter for use in evaluation and validation of high throughput screening assays. J Biomol Screen 4:67–73
16. Phillips C, Thrower J, Pickart CM, Hill CP (2001) Structure of a new crystal form of tetra-ubiquitin. Acta Crystallogr D Biol Crystallogr 57(Pt 2):341–344

Chapter 8

DNA Damage Response in *Xenopus laevis* Cell-Free Extracts

Tomas Aparicio Casado and Jean Gautier

Abstract

The DNA damage response (DDR) is a coordinated cellular response to a variety of insults to the genome. DDR initiates the activation of cell cycle checkpoints preventing the propagation of damaged DNA followed by DNA repair, which are both critical in maintaining genome integrity. Several model systems have been developed to study the mechanisms and complexity of checkpoint function. Here we describe the application of cell-free extracts derived from *Xenopus* eggs as a model system to investigate signaling from DNA damage, modulation of DNA replication, checkpoint activation, and ultimately DNA repair. We outline the preparation of cell-free extracts, DNA substrates, and their subsequent use in assays aimed at understanding the cellular response to DNA damage. Cell-free extracts derived from the eggs of *Xenopus laevis* remain a robust and versatile system to decipher the biochemical steps underlying this essential characteristic of all cells, critical for genome stability.

Key words *Xenopus* cell-free extracts, DNA replication, DNA damage response, Chromatin, S-phase, DNA repair

1 Introduction

Upon DNA damage, cells activate a DDR that causes a transient cell cycle arrest to facilitate DNA repair and/or programmed cell death. The DDR entails several steps: (1) recognition and signaling from the lesions, activation of cell cycle checkpoints, and fixing the lesions through a variety of DNA repair pathways. An operational DDR is critical for the maintenance of genome stability [1, 2].

Defects in sensing DNA damage, signaling from DNA damage, activating checkpoints, or repairing the lesions all lead to persistent DNA damage as well as improper propagation of genetic errors resulting in genomic instability and subsequent increased incidence in diseases, including cancer [3]. Genetic screens in yeast have been instrumental in identifying and characterizing components of the DDR, such as the identification of radiation-sensitive and checkpoint genes [4, 5]. These findings have been further validated in mammalian cells. However, the use of cellular systems has some limitations. For instance, standard genetic screens may be difficult to interpret for essential genes. Also, the DNA damage response is more complex in vertebrates than in yeast. Therefore, certain critical regulators of the damage response, such as p53 and BRCA1, can only be found in vertebrates [6]. Mammalian cell lines have been used to circumvent the above drawbacks. However, cell-based model systems do not allow the use of specific biochemical readouts because they are frequently based on cell growth, survival, or other phenotypes that derive from complex outputs [7]. Cell-free extracts from *Xenopus* provide an alternative that circumvents some of these limitations.

Xenopus cell-free extracts have been instrumental for the study of DNA replication, repair, and checkpoint signaling [8–16]. This system has key advantages [12, 17, 18]: First, the extracts contain cytoplasmic and nuclear proteins that can support up to 12 cell divisions in the absence of transcription. Second, the protein concentration in the extracts is high and sufficient to carry out DNA transactions, including complete rounds of semiconservative, cell cycle regulated DNA replication. Third, the extracts allow the study of essential proteins following immunodepletion. Indeed, downregulation or deletion of essential genes using siRNA or CRISPR leads to complex phenotypes due to the gradual decrease of protein expression, which varies depending on their half-life. Extracts allow biochemical rescue experiments with recombinant proteins, wild-type, or engineered to contain mutations or deletions to further dissect its function. For example, depletion of endogenous proteins from the extract using specific antibodies has allowed to study extensively the role of essential proteins in DNA replication, repair, and checkpoint signaling [8, 17, 19–22]. Additionally, the absence of zygotic transcription has allowed the study of proteins whose activity would be interfered by gene expression, such as the role of the oncogene c-Myc in DNA replication [23] or the replication and transcription-independent mechanism of interstrand cross-link repair [19, 24].

DNA replication in eukaryotes requires the sequential loading of replication initiation factors ORC1-6, Cdc6, and Cdt1-MCM2-7 that assemble on chromatin to form the prereplication complex. The subsequent initiation of DNA replication requires the activity of S-CDKs and DDK kinases and the recruitment of Cdc45 and GINS to form the active CMG helicase (Cdc45-MCM-GINS) as well as Treslin, MCM10, TopBP1, DNA Polymerase ϵ, and other replication factors before active replication forks are established [10]. DNA replication experiments with *Xenopus* extracts involve the addition of sperm chromatin to egg extracts made from unfertilized *Xenopus* eggs by crushing the eggs by centrifugation at low speeds ($\leq 20{,}000 \times g$) and recovering the fraction containing cytoplasmic and nucleoplasmic proteins along with lipids and small vesicles, with minimal dilution (Low-Speed Supernatant LSS, *see* Fig. 1). The haploid maternal genomic DNA is pelleted during

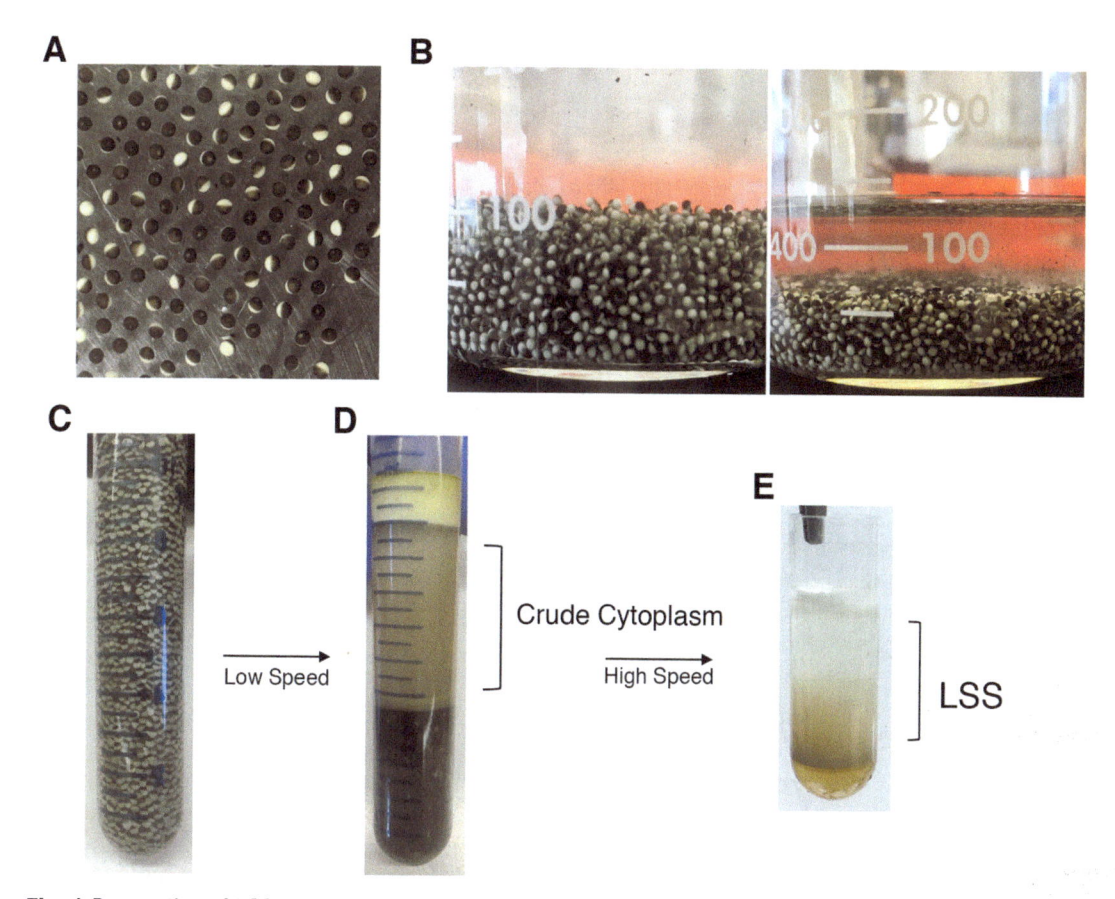

Fig. 1 Preparation of LSS and HSS extracts. (**a**) Freshly laid *Xenopus* eggs in their jelly coat. (**b**) Dejellied egg following incubation in a L-cysteine solution. (**c**) Packed eggs before low-speed centrifugation (**d**) After low-speed centrifugation crude cytoplasm is recovered by piercing the side of the tube with a needle. The crude cytosol is then further fractionated by ultracentrifugation. (**d**, **e**) clarified cytosol (transparent layer) and membranous layer are collected to form Low-Speed Supernatant Extract LSS. For HSS preparation, ultracentrifugation time is extended to separate the cytoplasm from lipid-rich layers better and only the clear layer (lacking membranes) is collected

extract crushing. LSS can assemble nuclei from demembranated sperm heads that are capable of undergoing one complete round of semiconservative DNA replication [8, 17]. By varying the method used to prepare the extract, different types of extracts mimicking specific phases of the cell cycle can be obtained (Fig. 2). These extracts have been particularly useful in understanding the regulation of entry into S phase and mitosis [25–30].

The initiation of DNA replication in extracts requires the assembly of a nuclear membrane surrounding the genomic DNA that locally imports and concentrates CDK and DDK kinase activities. Small DNA template, such as a circular plasmid, replicate poorly in crude cytosol (LSS), presumably because they do not promote nuclear membrane assembly efficiently. In contrast, extracts fractionated at high speeds ($\geq 200,000 \times g$, High-Speed

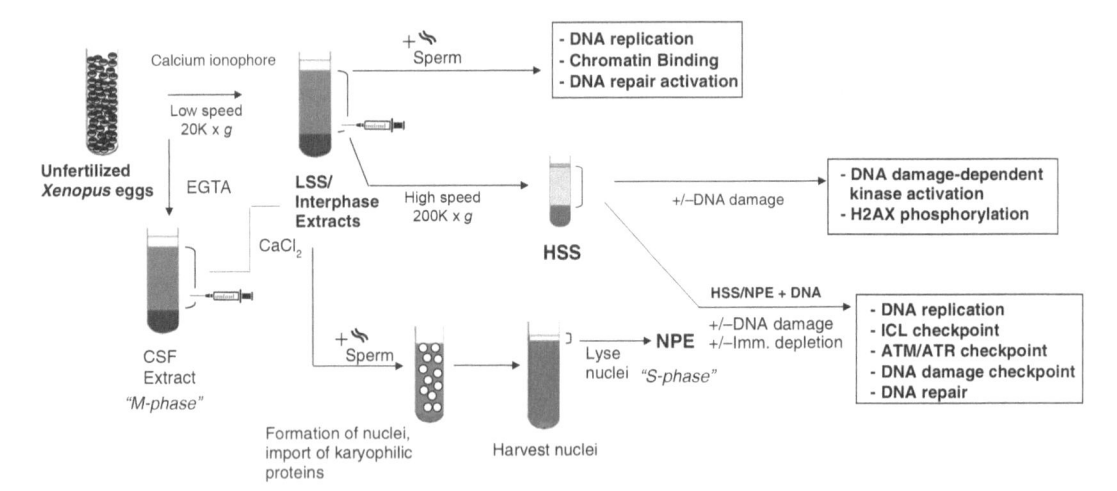

Fig. 2 Schematic representation of Xenopus extracts used to the study the DNA Damage Response in the different cell-cycle phases. Following distinct centrifugation steps, extracts can be made replication-competent (LSS for chromatin templates, HSS/NPE for small plasmid templates) or replication-incompetent (HSS). See the text for details

Supernatants, HSS) lack most lipids and vesicles and are unable to assemble nuclear envelopes, thus are replication-incompetent. HSS extracts are still capable of assembling chromatin and prereplication complexes on template DNA. Walter and Newport developed a "nucleus-free" system where the necessary kinase activity is supplemented in the form of an S-phase nucleoplasmic extract (NPE) derived from nuclei assembled in Low-Speed Supernatant extracts. Thus, artificially inducing the firing of prereplication complexes [22]. This allows for the study of plasmid replication and also alleviates other challenges faced in extracts involving nuclear assembly. Both types of *Xenopus* extracts ("nucleus-assembling" and "nucleus-free") can be used to study checkpoint signaling by experimentally interfering with DNA replication or introducing damaged DNA.

In this chapter, we describe techniques that use the *Xenopus* model system to recapitulate several aspects of the DNA damage response, including cell cycle checkpoints and some aspects of DNA repair. The protocols used in the preparation of various extracts are described in Subheading 3.1. The intact and damaged DNA templates used in conjunction with the extracts are described in Subheading 3.2. The subsequent section (Subheading 3.3) covers the various assays that involve the use of the *Xenopus* system to investigate cell cycle checkpoints.

2 Materials

The animals and hormones described below are materials common to the protocols used for the preparation of extracts.

2.1 Animals

Xenopus laevis (the African clawed frog), females and males (Nasco). Females are used for obtaining eggs for extract preparation for at least three times, while males are used in the preparation of demembranated sperm chromatin.

2.2 Hormones to Prime Frogs

Pregnant mare serum gonadotropin (PMSG, BioVendor) and pharmaceutical grade human chorionic gonadotropin HCG (Chorulon Injectable, Merk).

2.3 General Equipment

1. Refrigerated centrifuge (Sorvall RC-6 PLUS or similar) with HB-6 swinging bucket rotor and 14 mL round tube and Eppendorf tube adapters.
2. Refrigerated tabletop ultracentrifuge (Beckman OPTIMA MAX-XP or similar) equipped with TLS-55 swinging bucket rotor.
3. Fluorescent microscope (with a 385–400 nm excitation filter for DAPI-stained nuclei observation), plus glass slides and round 22 mm coverslips and hemocytometer.
4. Tabletop centrifuges for 14 mL and Eppendorf-type tubes.
5. Water bath or temperature-controlled incubator (e.g., Eppendorf Thermomixer) capable of maintaining a temperature of 21–21 °C.
6. Animal dissection kit.
7. Liquid nitrogen and cryogenic storage Dewar.
8. Glass beakers.
9. 2.5 mL ultracentrifuge tubes (Beckman Coulter Ultra-Clear # 347356).
10. 21-G needles and 3 mL disposable syringes.
11. 14 mL and 5 mL polypropylene Falcon tubes (Becton Dickinson # 352059 and #352063).
12. Low retention 1.5-mL Eppendorf tubes (Fisher Scientific).

2.4 CSF Extract

1. MMR (Marc's Modified Ringer's) buffer: 5 mM HEPES (pH 7.7–7.8), 0.1 mM EDTA, 100 mM NaCl, 2 mM KCl, 1 mM $MgCl_2$, and 2 mM $CaCl_2$.
2. Dejellying buffer: 2% L-cysteine hydrochloride monohydrate in MMR (pH 7.8 is critical).
3. XB Buffer: 10 mM HEPES (pH 7.7–7.8), 1 mM $MgCl_2$, 0.1 mM $CaCl_2$, and 100 mM KCl, 50 mM sucrose.

4. Cytostatic factor (CSF)-XB Buffer: 10 mM HEPES (pH 7.7–7.8), 2 mM MgCl$_2$, 0.1 mM CaCl$_2$, 100 mM KCl, 5 mM EGTA, and 50 mM sucrose.
5. Cytochalasin B. 10 mg/mL in ethanol. Store at −20 °C.
6. Energy Mix: 150 mM creatine phosphate, 20 mM ATP, 20 mM MgCl$_2$ (Store at −20 °C).
7. LPC protease inhibitors: leupeptin, pepstatin, and chymostatin (Roche). Make a stock with 10 mg/mL each in DMSO (1000× stock) and store at −20 °C.
8. 1 M CaCl$_2$, for release into Interphase.

2.5 Activated CSF Extract/Interphase Extract I

In addition to the materials required for CSF extract (*see* Subheading 2.1), the activated extract requires:

1. 10 mg/mL cycloheximide.
2. 1 M CaCl$_2$.

2.6 ELB Extract/LSS Extract

The materials required for the preparation of this extract are the same as those needed for CSF extract (*see* Subheading 2.1) except that ELB buffer is used instead of CSF buffer

1. Egg Lysis Buffer (ELB): 250 mM sucrose, 1× ELB Salts, 1 mM DTT, and 50 μg/mL cycloheximide.
2. 10× ELB Salts (for ELB buffer): 25 mM MgCl$_2$, 0.5 M KCl, and 100 mM HEPES, pH 7.7–7.8, with KOH; filter-sterilize. Store at 4 °C.
3. DAPI fix (for sperm decondensation and nuclei assembly observation): 3.7% formaldehyde, 2 μg/mL DAPI, 80 mM KCl, 15 mM NaCl, and 15 mM PIPES–KOH (pH 7.2), 50% glycerol, stored at −20 °C.

2.7 Interphase Extract II

1. Dejellying buffer: 2% L-cysteine hydrochloride monohydrate in 0.25× MMR, pH 7.8.
2. 5× MMR: 100 mM HEPES–KOH, pH 7.5, 2 M NaCl, 10 mM KCl, 5 mM MgSO$_4$, 10 mM CaCl$_2$, and 0.5 mM EDTA. Dilute to 1× or 0.25× as needed.
3. 5× S buffer: 250 mM HEPES–KOH, pH 7.5, 250 mM KCl, 12.5 mM MgCl$_2$, and 1.25 M sucrose. Keep at 4 °C. Add β-mercaptoethanol to a final concentration of 2 mM; and LPC protease inhibitors (optional).
4. DAPI fix (for sperm decondensation observation): 3.7% formaldehyde, 2 μg/mL DAPI, 80 mM KCl, 15 mM NaCl, 15 mM PIPES–KOH (pH 7.2), and 50% glycerol, stored at −20 °C.

Store the following reagents in small aliquots at −20 °C:

DNA Damage Response in Cell-Free Extracts   109

(a) Calcium Ionophore A23187 (Sigma). Stock: 10 mg/mL in DMSO.

(b) 10 mg/mL Cytochalasin B (Sigma) in DMSO or ethanol.

(c) 1 M creatine phosphate (Sigma) in Milli-Q water.

(d) 10 mg/mL creatine phosphokinase (CPK, Sigma) in Milli-Q water.

(e) 15 mg/mL leupeptin (Roche) in DMSO.

(f) 10 mg/mL cycloheximide (Sigma) in Milli-Q water.

(g) Other materials: Styrofoam box or wooden platform, liquid nitrogen, cryogenic vials (1.5- or 2-mL).

2.8 High-Speed Supernatant (HSS)

1. Marc's Modified Ringer's Buffer (MMR): 100 mM NaCl, 2 mM KCl, 0.5 mM MgSO$_4$, 2.5 mM CaCl$_2$, 0.1 mM EDTA, and 5 mM HEPES–KOH, pH 7.7–7.8.

2. Dejellying solution: 2% L-cysteine hydrochloride monohydrate in MMR, pH 7.8.

3. Egg Lysis Buffer (ELB): 1× ELB salts, 1 mM DTT, 50 µg/mL cycloheximide, and 0.25 M sucrose.
   The 10× salts (stock) for Egg Lysis Buffer: 25 mM MgCl$_2$, 500 mM KCl, and 100 mM HEPES–KOH, pH 7.8, filter-sterilized and stored at 4 °C.

4. Cytochalasin B (Sigma) 5 mg/mL in ethanol; stored at −20 °C.

5. Cycloheximide (Sigma) 10 mg/mL in water and stored at −20 °C.

6. Dithiothreitol (DTT) 1 M in water; stored at −20 °C.

7. LPC protease inhibitors: leupeptin, pepstatin, and chymostatin. Make a stock with 10 mg/mL each in DMSO (1000× stock) and store at −20 °C.

8. Energy Mix: 10 mM creatine phosphate, 10 µg/mL creatine kinase, 2 mM ATP, 2 mM MgCl$_2$, 5 mM HEPES, pH 7.5, and 1 mM DTT.

2.9 Nucleoplasmic Extract (NPE)

1. Marc's Modified Ringer's Buffer (MMR): 100 mM NaCl, 2 mM KCl, 0.5 mM MgSO$_4$, 2.5 mM CaCl$_2$, 0.1 mM EDTA, and 5 mM HEPES–KOH, pH 7.7–7.8.

2. Dejellying solution: 2% L-cysteine hydrochloride monohydrate in MMR, pH 7.8.

3. Egg lysis buffer: 250 mM sucrose, 2.5 mM MgCl$_2$, 50 mM KCl, 10 mM HEPES, pH 7.8, 50 µg/mL cycloheximide, 1 mM DTT.

4. LPC protease inhibitor cocktail: leupeptin, pepstatin, and chymostatin. Make a stock with 10 mg/mL each in DMSO (1000× stock) and store at −20 °C.

5. Cytochalasin B (Sigma) 5 mg/mL in ethanol; stored at −20 °C.

6. Cycloheximide (Sigma) 10 mg/mL in water and stored at −20 °C.

7. Nocodazole (Sigma) 5 mg/mL in DMSO; Sigma.

8. Dithiothreitol (DTT) 1 M in water; stored at −20 °C.

9. NPE–ATP regeneration system: Prepare stocks of 0.2 M ATP (pH 7.0), 1 M phosphocreatine disodium salt (pH 7.0) and 5 mg/mL creatine phosphokinase (CPK) in in 50 mM NaCl, 50% glycerol, and 10 mM HEPES–KOH (pH 7.5). Make small aliquots and store at −20 °C for up to six months. Make fresh before use. Mix on ice 10 μL of phosphocreatine, 5 μL of 0.2 M ATP, and 0.5 μL of 5 mg/mL CPK. Use 1 μL per 30 μL of extract. Discard leftovers after use.

10. Demembranated Sperm Chromatin (*see* Subheadings 2.8 and 3.2.1).

11. DAPI fix (for nuclear assembly observation): 3.7% formaldehyde, 2 μg/mL DAPI, 80 mM KCl, 15 mM NaCl, 15 mM PIPES–KOH (pH 7.2), and 50% glycerol, stored at −20 °C.

12. Falcon Tube (Becton Dickinson # 352063).

13. Becton Dickinson tube # 352059.

14. Beckman tube # P60720.

2.10 Immunodepletion of Extracts to Study Protein Function

1. Protein-A Sepharose CL-4B agarose beads (GE Healthcare).
2. Spin Columns (Snap Cap, Pierce #69725 or equivalent).
3. Phosphate-buffered saline (PBS buffer), pH 7–7.5.

2.11 Demembranated Sperm Chromatin

1. Benzocaine. Dissolve 1 g in 10 mL of 100% ethanol. Dispense dropwise into 2 L of water with continuous stirring. This gives a 0.05% solution of benzocaine. Make fresh. IMPORTANT: always follow your Institutional Animal Care and Use Committee guidelines on animal euthanasia.

2. Marc's Modified Ringer's Buffer (MMR): 100 mM NaCl, 2 mM KCl, 0.5 mM MgSO$_4$, 2.5 mM CaCl$_2$, 0.1 mM EDTA, and 5 mM HEPES–KOH, pH 7.7–7.8.

3. Nuclei Preparation Buffer (NPB): 250 mM sucrose, 15 mM HEPES–KOH, pH 7.8, 1 mM EDTA, 0.5 mM Spermidine, 0.2 mM Spermine, and 1 mM DTT. Prepare as a 2× solution.

4. Triton X-100, 10% v/v in water; store at room temperature.

5. Bovine serum albumin. 10% w/v in water. Store at −20 °C or make fresh.
6. Glycerol.
7. PMSF 0.3 M made in 100% ethanol. Make fresh before use.
8. LPC protease inhibitors: leupeptin, pepstatin, and chymostatin. Make a stock with 10 mg/mL each in DMSO (1000× stock) and store at −20 °C.
9. Spermidine 10 mM.
10. Spermine 10 mM.
11. Sucrose 1.5 M; store at 4 °C.

2.12 Generation of Double-Strand Breaks (DSB) in Chromatin

1. LSS or HSS *Xenopus* egg extract.
2. Demembranated sperm nuclei (*see* Subheading 3.2.1).
3. PflMI or equivalent restriction endonuclease.
4. Chromatin isolation buffer: 50 mM HEPES–KOH pH 7.5, 50 mM KCl, 2.5 mM $MgCl_2$, and 0.125% Triton X-100.
5. Sucrose cushion: 30% sucrose in Chromatin Isolation buffer without Triton X-100.

2.13 Preparation of plasmids containing Interstrand Cross-Links (ICLs)

1. pBS plasmid.
2. SJG-136 (NCI/Spriogen LTD).
3. Triethanolamine, 50 mM in EDTA, 2 mM.
4. Ethanol for DNA precipitation.

2.14 Preparation of DNA with Double-Strand Breaks

1. pBR322 plasmid.
2. HaeIII restriction endonuclease (NEB).
3. Phenol–chloroform to extract digested DNA.

2.15 Preparation of Biotinylated Substrates

1. DNA fragment of interest.
2. Gel purification kit.
3. T4 polymerase.
4. dATP, dGTP, dTTP, and biotin–dCTP.
5. EDTA, 0.5 M pH 8.0.
6. PCR purification kit (Qiagen).

2.16 DNA Damage Checkpoint Induced by Single-Strand DNA Gaps in Chromatin

1. LSS or HSS *Xenopus* egg extract.
2. Demembranated sperm nuclei.
3. Etoposide or Exonuclease III.
4. Tris–HCl, 1 M pH 8.0.
5. $MgCl_2$, 1 M.

6. Nuclei Preparation Buffer (NPB): 250 mM sucrose, 15 mM HEPES–KOH, pH 7.8, 1 mM EDTA, 0.5 mM spermidine, 0.2 mM spermine, 1 mM DTT, LPC protease inhibitors 10 μg/mL each, and 0.3 mM PMSF.
7. Caffeine 100 mM in PIPES 10 mM, pH 8.0. Prepare fresh before use.
8. Specific PIKK inhibitors (KU55933 for ATM and VE-821 for ATR, Selleck Chemicals Inc.).
9. α-^{32}P-dCTP (PerkinElmer).
10. Stop Solution: 0.5% SDS, 80 mM Tris (pH 8.0), and 8 mM EDTA.
11. Proteinase K solution (Roche).
12. Phenol–chloroform to extract DNA.
13. Trichloroacetic acid, 50% w/v in water.
14. Whatman 3 MM blotting paper.

2.17 DNA Damage Checkpoint Induced by DSB in Chromatin

1. LSS or HSS *Xenopus* egg extract.
2. Circular control plasmid or plasmids containing double-strand breaks (DSB).
3. Caffeine, 100 mM in PIPES, pH 8.0. Prepare fresh before use.
4. Specific PIKK inhibitors (KU55933 for ATM and VE-821 for ATR, Selleck Chemicals Inc).
5. Sodium Acetate, 3 M.
6. Stop Solution: 0.5% SDS, 80 mM Tris (pH 8.0), and 8 mM EDTA.
7. Proteinase K solution (Roche).
8. Phenol–chloroform to extract DNA.
9. Trichloroacetic acid.
10. Whatman 3 MM blotting paper.

2.18 Assay to Study Checkpoint Induced by ICLs

1. *Xenopus* extracts (HSS/NPE, Subheadings 3.1.5 and 3.1.6).
2. Control and ICL containing DNA (Subheading 3.2.3).
3. NPE–ATP regeneration system: Prepare stocks of 0.2 M ATP (pH 7.0), 1 M phosphocreatine disodium salt (pH 7.0), and 5 mg/mL creatine phosphokinase (CPK) in 50 mM NaCl, 50% glycerol, 10 mM HEPES–KOH (pH 7.5). Make small aliquots and store at −20 °C for up to 6 months. Make fresh before use. Mix on ice 10 μL of phosphocreatine, 5 μL of 0.2 M ATP, and 0.5 μL of 5 mg/mL CPK. Use 1 μL per 30 μL of extract. Discard leftovers after use.
4. Caffeine, 100 mM in PIPES, pH 8.0. Prepare fresh before use.

5. Specific PIKK inhibitors (KU55933 for ATM and VE-821 for ATR, Selleck Chemicals Inc).
6. α-^{32}P-dCTP (PerkinElmer).
7. Stop Solution: 0.5% SDS, 80 mM Tris (pH 8.0), 8 mM EDTA-Proteinase K solution (Roche).
8. Phenol–chloroform–isoamyl alcohol (25:24:1, Sigma).
9. Sodium acetate, 3 M.
10. Ethanol.
11. DNA loading buffer with bromophenol blue and cyan blue dyes.
12. Trichloroacetic acid.

2.19 Study of the Checkpoint Induced by DNA Damage In Trans on DNA Replication

1. *Xenopus* Extracts (HSS/NPE; *see* Subheadings 3.1.5 and 3.1.6).
2. NPE–ATP regeneration system: Prepare stocks of 0.2 M ATP (pH 7.0), 1 M phosphocreatine disodium salt (pH 7.0) and 5 mg/mL creatine phosphokinase (CPK) in in 50 mM NaCl, 50% glycerol, 10 mM HEPES–KOH (pH 7.5). Make small aliquots and store at −20 °C for up to six months. Make fresh before use. Mix on ice 10 μL of phosphocreatine, 5 μL of 0.2 M ATP, and 0.5 μL of 5 mg/mL CPK. Use 1 μL per 30 μL of extract. Discard leftovers after use.
3. Control or ICL DNA (*see* Subheading 3.2.3).
4. α-^{32}P-dCTP (PerkinElmer).

2.20 Phosphorylated Histone H2AX Detection (Endogenous)

1. Activated CSF extract or Interphase Extract I (*see* Subheading 3.1.2).
2. Demembranated sperm chromatin (*see* Subheading 3.2.1).
3. Chromatin isolation buffer: 50 mM HEPES–KOH pH 7.5, 50 mM KCl, 2.5 mM MgCl$_2$, LPC protease inhibitors, 0.125% Triton X-100 (made fresh before use); this buffer is supplemented with 1 mM NaF, 1 mM Na vanadate, and 0.125% Triton X-100.
4. Sucrose cushion: Chromatin isolation buffer (without Triton X-100) supplemented with 30% sucrose.

2.21 Histone H2AX Phosphorylation Assay (Exogenous Substrate)

1. *Xenopus* extracts (*see* Subheading 3.1).
2. Biotinylated DNA fragments to induce DNA damage.
3. Streptavidin magnetic beads (Dynabeads, Thermo Fisher).
4. ELB Buffer: 250 mM sucrose, 1× ELB Salts, 1 mM DTT, 50 μg/mL cycloheximide (10× ELB Salts—for ELB buffer: 25 mM MgCl$_2$, 0.5 M KCl, 100 mM HEPES, pH 7.7–7.8 with KOH); filter-sterilize. Store at 4 °C.

5. Wild-type H2AX peptide: AVGKKASQASQEY.
6. Mutant H2AX peptide: AVGKKAAQAAQEY.
7. EB Buffer: 20 mM HEPES, pH 7.5, 50 mM NaCl, 10 mM MgCl$_2$, 1 mM DTT.
8. ATP.
9. γ-32P-ATP (PerkinElmer).
10. EDTA, 0.5 M pH 8.0.
11. p81 phosphocellulose filter papers (Upstate Biotechnology).
12. Acetic acid.
13. Caffeine, 100 mM in PIPES, pH 8.0. Prepare fresh before use.
14. Specific PIKK inhibitors (KU55933 for ATM and VE-821 for ATR, Selleck Chemicals Inc).
15. EB Kinase buffer: 20 mM HEPES, pH 7.5, 50 mM NaCl, 10 mM MgCl$_2$, 1 mM DTT, 1 mM NaF, 1 mM Na$_3$VO$_4$, 10 mM MnCl$_2$.

2.22 Phosphorylation of ATM/ATR Target Proteins

1. *Xenopus* Extracts (*see* Subheading 3.1).
2. Energy Mix: 10 mM creatine phosphate, 10 μg/mL creatine kinase, 2 mM ATP, 2 mM MgCl$_2$, 5 mM HEPES, pH 7.5, 1 mM DTT.
3. Control and damaged DNA (either chromatin or small DNA templates can be used; *see* Subheading 3.2).
4. Caffeine, 100 mM in PIPES, pH 8.0. Prepare fresh before use.
5. Specific PIKK inhibitors (KU55933 for ATM and VE-821 for ATR, Selleck Chemicals Inc).
6. Phospho-Chk1 antibody (Cell Signaling).
7. Phospho-ATM antibody (Rockland Immunochemicals, PA).
8. Geminin, roscovitine, or aphidicolin for DNA replication inhibition.
9. Curcumin dissolved in ethanol.

2.23 Chromatin Binding Assay

1. *Xenopus* extracts (*see* Subheading 3.1).
2. Demembranated sperm chromatin (*see* Subheading 3.2.1).
3. Chromatin isolation buffer: 50 mM HEPES–KOH pH 7.5, 50 mM KCl, 2.5 mM MgCl$_2$, 0.125% Triton X-100 (made fresh before use).
4. Chromatin isolation buffer + Sucrose: 50 mM HEPES–KOH pH 7.5, 50 mM KCl, 2.5 mM MgCl$_2$, 30% sucrose.
5. Micrococcal Nuclease (NEB).

2.24 Binding Assay Using HSS/NPE and ICL Plasmid

1. *Xenopus* extracts (HSS/NPE *see* Subheadings 3.1.5 and 3.1.6).
2. NPE–ATP regeneration system: Prepare stocks of 0.2 M ATP (pH 7.0), 1 M phosphocreatine disodium salt (pH 7.0) and 5 mg/mL creatine phosphokinase (CPK) in in 50 mM NaCl, 50% glycerol, 10 mM HEPES–KOH (pH 7.5). Make small aliquots and store at −20 °C for up to 6 months. Make fresh before use. Mix on ice 10 µL of phosphocreatine, 5 µL of 0.2 M ATP, and 0.5 µL of 5 mg/mL CPK. Use 1 µL per 30 µL of extract. Discard leftovers after use.
3. Protein-A Sepharose CL-4B beads.
4. Spin Columns (Snap Cap, Pierce #69705 or equivalent).
5. Phosphate-buffered saline (PBS buffer), pH 7–7.5.
6. Control or ICL plasmid DNA (*see* Subheading 3.2.3).
7. Chromatin isolation buffer (high KCL): 50 mM HEPES–KOH pH 7.5, 100 mM KCl, 2.5 mM $MgCl_2$, supplement with 0.125% Triton X-100.
8. Sucrose cushion: 30% sucrose in chromatin isolation buffer (without Triton X-100).

3 Methods

The methods described below include (**a**) the preparation of different types of extracts (Subheading 3.1), (**b**) the preparation of DNA substrates—both chromosomal (Subheadings 3.2.1 and 3.2.2) and small template DNA (Subheadings 3.2.3–3.2.5) and (**c**) assays used to study checkpoints—(1) DNA replication assays (Subheadings 3.3.1 and 3.3.2), (2) assays to study protein modifications such as phosphorylation to detect checkpoint activation (Subheadings 3.3.5–3.3.8), and (3) Chromatin–DNA binding assays (Subheadings 3.3.9 and 3.3.10).

3.1 Preparation of Extracts

Studies of the DDR using *Xenopus* involve at least four different types of extracts. Three of the extracts (Cytostatic Factor Extract, Low-Speed Interphase Extract, and High-Speed Extract) are derived from *Xenopus* eggs, whereas the fourth type (Nucleoplasmic Extract) requires *Xenopus* eggs and sperm chromatin to assemble and isolate nuclei from which a nucleoplasm fraction is extracted. The addition of sperm chromatin to activated CSF or LSS extracts leads to nuclear assembly and chromosomal replication. HSS lacks lipid vesicles and is unable to form nuclear envelopes; therefore, it is not replication competent, although DNA templates incubated in HSS form prereplication complexes. DNA replication can be initiated subsequentially in HSS by supplementing with NPE (Fig. 2) [16, 22].

Below, we describe the preparation of cytostatic-arrested (CSF) extracts arrested in M phase and activated CSF extracts. Since CSF and activated CSF extracts are obtained by relatively low-speed centrifugation, they are also referred to as Low-Speed Supernatant (LSS, Fig. 1). These extracts can be further fractionated for checkpoint studies to obtain fractionated membrane-free cytosol. Of note, a modification of the protocol to prepare LSS allows storing the extract, without appreciable loss of activity. We are referring to such extract as Interphase Extract II, which is also described in this chapter. Furthermore, we describe the preparation of High-Speed Supernatant (HSS), which is a membrane-free extract that supports the assembly of the prereplication complexes on chromatin [16, 22]. HSS extract is also used with Nucleoplasmic Extract (NPE) to bypass the nuclear envelop formation step, thus allowing normal DNA replication of chromatin and plasmid [16, 22, 31]. The NPE system is particularly useful since it allows for the modification of the nuclear environment and supports plasmid replication very efficiently [22, 32]. The preparation of demembranated sperm chromatin and NPE are also outlined in this chapter. Energy Mix is added to extracts before use. Figure 2 summarizes the various types of extracts and their applications.

3.1.1 CSF Extract

Mature unfertilized eggs laid by *Xenopus* are arrested in meiosis by a calcium-sensitive activity (Cyclo-Static Factor, CSF) that inhibits the release from metaphase II into anaphase and meiotic exit into interphase. In *Xenopus* CSF extracts, the mitotic state stabilized by CSF is preserved by chelating free calcium during extract preparation, thus maintaining the characteristics of an M phase of the cell cycle. The addition of calcium allows controlled arrest release into interphase. CSF extract has been used extensively over the years for reconstitution of cell cycle transitions, studies of chromosome condensation, cohesion and microtubule spindle assembly [33]. Extracts are typically prepared in the presence of cytochalasin B or D to prevent actin polymerization, preventing local gelation that impairs extract homogeneity and might affect other processes. Protocols for the preparation of extracts in the absence of actin polymerization inhibitors are available [34]. The preparation of fresh CSF-arrested extracts by our group has been previously described in [7, 35].

1. Prime 4–6 female frogs with 500 μL/frog of PMSG (PMSG stock: 100 IU/μL). This step is generally performed 3–7 days before the frogs are induced with HCG.

2. Induce females to lay eggs with 800 IU of HCG/frog injected the evening before egg collection. Place each frog in a container with at least 1 L of 1× MMR (pH 7.7–7.8) after HCG injection. Eggs are laid overnight and collected the next morning (Fig. 1a, *see* **Notes 1** and **2**).

3. Wash eggs in 1× MMR. Discard bad eggs with a plastic Pasteur pipette. Remove as much buffer as possible by decanting. Transfer eggs to a glass beaker.

4. Dejelly the eggs in 2% L-cysteine pH 7.8 (Fig. 1b, *see* **Note 3**).

5. Remove the cysteine solution by pouring off the excess liquid.

6. Rapidly wash eggs three times with XB and then remove all XB.

7. Wash eggs three times in CSF-XB and remove as much buffer as possible.

8. Wash eggs two times in XB containing protease inhibitors (LPC—leupeptin, pepstatin, and chymostatin, 1000× stock: 10 mg/mL each in DMSO).

9. Transfer eggs into 1 mL of CSF-XB with LPC protease inhibitors and 100 μg/mL cytochalasin B [36] in 1.5 mL Eppendorf tubes (*see* **Note 4**). Alternatively, if there is a large volume of eggs, 14-mL Falcon tubes can be used.

10. Pack the eggs by spinning for 1 min at 800 rpm (100 × g) and remove excess buffer (Fig. 1c, *see* **Note 5**).

11. Combine eggs from different tubes and fill each Eppendorf tube as much as possible. Crush the eggs at 4–8 °C for 15 min at 10,000 rpm (16,000 × g).

12. Collect the extract with an 18-gauge needle by puncturing the side of the tube and gently aspirate out the cloudy intermediate cytoplasmic layer. This layer is located between the top opaque lipid layer (yellow) and the solid pellet of pigments and egg debris (dark brown).

13. Supplement the cytosolic extract with 1/1000 volume of cytochalasin B (10 mg/mL stock), 1/15 volume of energy mix, LPC protease inhibitors, and 1/40 volume of 2 M sucrose.

14. Clarify the cytosolic extract by centrifugation for 30 min at 13,000 × g in an Eppendorf centrifuge at 4 °C. The transparent cytoplasmic layer is collected for immediate use (*see* **Notes 6 and 7**).

3.1.2 Activated CSF Extract/Interphase Extract I

1. Upon activation by calcium, CSF extracts exit M phase and enter interphase. Such activated extracts are used in DNA replication, chromatin binding, and checkpoint assays.

2. Supplement freshly prepared CSF extracts (*see* Subheading 3.1.1) with 50 μg/mL of cycloheximide and 0.4 mM $CaCl_2$.

3. Incubate for 15 min at 21 °C. This treatment mimics the Calcium flux that usually takes place after fertilization of eggs and triggers degradation of mitotic cyclins, inactivation of Cdc2/Cyclin B, and exit from mitosis (*see* **Note 7**).

3.1.3 ELB Extract/LSS Extract

A variation of the activated extract is to prepare a "crude" S-phase extract/Low-Speed Supernatant Extract with MMR buffer that does not contain EGTA. Thus, eggs are already activated, and the buffer does not need to be supplemented with $CaCl_2$. The protocol, in this case, is the same as for CSF extract except that the eggs are washed with ELB (Egg Lysis Buffer) instead of CSF-XB buffer. Since the ELB buffer already contains sucrose, it is also not necessary to supplement the cytosolic extract with 1/40 volume of 2 M sucrose (during **step 13** of CSF extract Subheading 3.1.1). Supplement extract with energy mix before use.

3.1.4 Interphase Extract II

The major advantage of this type of extract is that it can be stored as frozen drops at −80 °C and thawed before use [37], allowing multiple experiments or technical replicates to be performed with a batch of extracts on different days. This particularly useful when working with extracts depleted with precious antibodies. Frozen extracts can be kept for about six months without appreciable loss of quality.

1. Induce 4–6 frogs to lay eggs in 0.25× Interphase Extract II MMR (*see* MMR recipe, the frogs are induced to lay eggs as described in Subheading 3.1.1 above.

2. Collect and pool good eggs by aspirating them out with a plastic transfer pipette.

3. Decant as much buffer as you can and then add dejelly buffer (prepare 100 mL buffer per frog), for 5–10 min. If there are more eggs, change dejelly buffer every 2–3 min, but extending the time of dejellying may be detrimental to the quality of the extract (*see* **Note 3**).

4. Wash three times with 0.25× MMR buffer.

5. Add 1/10,000 of 10 mg/mL Calcium ionophore A23187 and incubate for 5 min to activate eggs. You might see how egg hemispheres (poles) become more apparent.

6. Wash three times with 0.25× MMR buffer.

7. Wash three times with S-buffer. Prepare 50 mL per frog, add + 7 μL β-mercaptoethanol per 50 mL, and in the last wash, add 1/1000 leupeptin.

8. Transfer eggs to a 14-mL Falcon tube.

9. Spin in Sorvall swing-bucket HB-6 rotor for 1 min at 800 rpm (100 × *g*). This is the packing spin. Remove excess buffer.

10. Spin again at 10,000 rpm (16,000 × *g*) for 30 min at 4 °C. This is the crushing spin (Fig. 1d).

11. Pool and transfer the cytoplasm layer to a 5 mL Falcon tube using a 21-G needle/syringe. Care should be taken not to aspirate the dark layer immediately below containing mitochondria.

12. Add cytochalasin B (1/500).

13. Seal tubes tightly with Parafilm and rotate in the cold room for 5 min to completely homogenize the extract.

14. Centrifuge in Beckman tabletop centrifuge with TLS-55 swing-bucket rotor for 55,000 rpm (~250,000 × g) for 15 min at 4 °C.

15. Take only the cytoplasmic and membrane layers. Leave out the mitochondrial layer, the oily layer and the debris (typically, the debris accumulates at the bottom-most part of the tube; immediately above the debris is the oily layer above which is the mitochondrial layer). Use a cut pipette tip (P1000 or P200) to recover the desired fractions (Fig. 1e).

16. Supplement with the following:

1/30	30 mM Creatine phosphate
1/60	150 µg/mL Phosphocreatine kinase
1/500	20 µg/mL Cycloheximide

17. Invert tube several times to make sure all components are adequately mixed.

18. Use extract within 3 h or freeze as follows

    Freezing the Interphase Extract II:

    (a) Add glycerol (3% final concentration) to extract and mix thoroughly by inverting the tube.

    (b) On a Styrofoam platform, place P60 plastic petri dishes filled with liquid nitrogen. Fill labeled cryotubes with liquid nitrogen and place on Styrofoam rack.

    (c) With a cut P200 tip, drop beads of 25 µL of extract into a P60 dish containing liquid nitrogen. Allow extract to flash-freeze in the liquid nitrogen (*see* **Note 8**).

    (d) Store the beads of frozen extract in cryogenic vials at −80 °C.

3.1.5 High-Speed Supernatant (HSS)

1. Prime 4–6 female frogs with 500 µL/frog of PMSG (PMSG stock: 100 IU/µL) within a week of extract preparation.

2. Induce females to lay eggs with 800 IU of HCG/frog injected the night before egg collection. Place each frog in a container with 1 L of 1× MMR (pH 7.7–7.8) after HCG injection (*see* **Note 9**). Eggs are laid overnight and collected the next morning.

3. Wash eggs with 1× MMR. Remove dead eggs/floaters/debris with a transfer pipette.

4. Combine eggs in a glass beaker and dejelly the eggs using a 2% L-cysteine solution (pH 7.8) for about 5–10 min (*see* **Note 3**).

5. Wash eggs three times with 0.5× MMR, pH 7.8 (*see* **Note 10**).

6. Wash eggs three times with ELB, pH 7.8. At the last wash, add 1 mL of 10 mg/mL cycloheximide per 100 mL ELB.

7. Transfer eggs into 14-mL polypropylene tubes (Falcon tube #352059) without caps and centrifuge them at 800 rpm (100 × g) for 1 min at 4 °C (Sorvall HB-6 rotor, use adapters to enable centrifugation of the tubes). This is the packing spin. Remove excess buffer after this spin and add cytochalasin B on top to a final concentration of 5 µg/mL (stock 10 mg/mL).

8. Centrifuge eggs at 10,000 rpm (16,000 × g) for 20 min at 4 °C. This is the crushing spin and will result in the formation of different layers.

9. Transfer the cytoplasmic layer (central layer) to a 50-mL tube. Remove the layer by using a 5-mL syringe with a 25-gauge needle. Add 100 µg/mL cycloheximide, 5 µg/mL cytochalasin B, 1 mM DTT, and LPC protease inhibitors.

10. Pipet cytoplasmic layer in ultracentrifuge tubes (make sure to have at least 1.75 mL per tube or the tube might be crushed onto itself). Centrifuge tubes at 55,000 rpm for 2 h at 4 °C (TLS-55 rotor ~250,000 × g, *see* **Note 11**).

11. Collect clear cytoplasm with a pipette with a cutoff tip, avoiding the membrane rich layer below. You might respin again for 15 min at 55,000 rpm if required.

12. Aliquot into 0.5 mL tubes (100 µL/tube), freeze with liquid nitrogen, and store at −80 °C. HSS is stable for several years.

3.1.6 Nucleoplasmic Extract (NPE)

The preparation of nucleoplasmic extracts was initially reported by Walter et al. [22]. Nucleoplasmic extracts, when used in combination with HSS, have several advantages. In particular, they support plasmid replication [16, 21, 22, 38]. Therefore, this type of extract allows to study modified plasmid DNA templates to study the DNA damage response [19, 24].

1. Inject 20 female frogs with PMSG (500 µL/frog). Leave them for 3–5 days.

2. Induce egg-laying by injecting each frog with 1 mL of HCG and placing them in individual containers with 1 L 1× MMR. It must be noted that the final yield of NPE will be a small fraction of the crude S extract obtained from the eggs (*see* **Notes 1** and **2**).

3. Combine good eggs, remove and discard buffer. Wash eggs with 1× MMR.
4. Dejelly the eggs with 2% L-cysteine pH 7.8 (*see* **Note 3**).
5. Wash the eggs rapidly three times with 0.5× MMR and three times with 1× ELB, pH 7.8. Transfer eggs to 14-mL Falcon Tube (# 352059).
6. Pack eggs by centrifugation for 1 min at 800 rpm (100 × g). Remove excess buffer, add LPC protease inhibitor cocktail, Aprotinin 5 µg/mL (1:2000), and 2.5 µg/mL cytochalasin B.
7. Crush eggs by centrifugation at 10,000 rpm (16,000 × g) in a Sorvall Centrifuge with an HB6 rotor for 15 min at 4 °C.
8. The cytoplasmic layer is withdrawn by puncturing the side of the tube with a 21-gauge needle. This is the crude S extract (the middle layer), and is supplemented with LPC protease inhibitor cocktail, 10 µg/mL Aprotinin, 5 µg/mL Cytochalasin B, 50 µg/mL cycloheximide, 3.3 µg/mL nocodazole (stock 1000× in DMSO), and 1 mM DTT.
9. The crude S-phase extract thus obtained is recentrifuged at 10,000 rpm for 10 min at 4 °C in a Falcon tube of appropriate size.
10. All of the residual lipids and the viscous dark brown material located just below it are entirely aspirated off. About 1 mL of the extract could be lost at this step, but all of the upper layers must be removed. The remaining cytoplasm is decanted into a fresh tube and supplemented with NPE–ATP regeneration system (Subheading 2.6, prepare fresh before use).
11. Add demembranated sperm chromatin (*see* Subheading 3.2.1). This nuclear assembly reaction is carried out in volumes of 4.5 mL per tube. The assembly reaction is made by thoroughly mixing 1 mL of the extract with 90 µL of 200,000/µL sperm. Pipetting several times with a cut blue tip is recommended before transferring the mix into a 5-mL Falcon tube (Becton Dickinson # 352063). To this mix, 3.5 mL of the extract is added and mixed by inverting multiple times. It is essential to have at least 4 mL of extract in each tube to ascertain that the layer of nuclei formed on the top after centrifugation will be thick enough to harvest.
12. The nuclear assembly reaction is incubated at room temperature and mixed by inversion every 10 min. A few microliters of the reaction are observed under the microscope. It is recommended to wait approximately 30 min after the time when round nuclei first become visible. The total time after sperm addition is around 70–80 min. Large nuclei are essential to ensure a good layer of nuclei in the subsequent step.

13. To collect the assembled nuclei, the reaction is centrifuged for 2 min at 10,000 rpm (16,000 × g) at 4 °C in an HB-4 rotor in a Sorvall centrifuge. The Falcon tube (Becton Dickinson # 352063) with the extract is placed in a tube (Becton Dickinson # 352059) containing 2 mL of water to help support it during the spin. After this step, the nuclei should be visible as a clear layer about 2 mm thick on the top of the extract. If the layer is thinner than expected, let the nuclei grow more next time and make sure that the sperm concentration used is correct. Nocodazole must be added at this step since it prevents microtubule assembly allowing nuclei to float to the top of the tube during centrifugation.

14. Hold the tube to the light to better distinguish the clear nuclear layer from the underlying cytoplasm and remove the nuclei with a cutoff 200 μL pipette tip. Transfer all of the nuclei to a Beckman 5 × 21 mm ultracentrifuge tube (# P60720).

15. Supplement the nuclei with ELB to a final volume of 225 μL (assuming a 4.5-mL nuclear assembly reaction); mix thoroughly with a cutoff pipette tip and centrifuge for 30 min at 55,000 rpm (250,000 × g) at 2 °C. Centrifugation is carried out in a Beckman TL100 tabletop ultracentrifuge using a TL55 swinging bucket rotor furnished with Teflon adapters (Beckman) to accommodate the 5 × 21 mm tubes. If there is more than 750 μL of the NPE + ELB mixture, a thick-walled 11 × 34 mm tube (Beckman # 343778) can be used with a centrifugation time of 40 min.

16. Carefully aspirate any lipids at the top of the sample after the spin and harvest the clear nucleoplasm taking care to avoid the pellet consisting of nuclear envelopes and chromatin (*see* **Notes 12** and **13**).

17. The expected volume of NPE from each 4.5 mL nuclear assembly reaction is around 180 μL. NPE is snap-frozen (in 10–20 μL aliquots) in liquid nitrogen, stored at −80 °C and is stable for several years if freeze–thawing is prevented.

3.1.7 Immunodepletion of Extracts to Study Protein Function

Quantitative immunodepletion allows the complete removal of an endogenous protein from the extract by using antibodies against the protein of interest. The removal of proteins or protein complexes from the extract is a valuable tool to investigate protein function; however, care should be taken to retain the overall activity and functionality of the extract. Therefore, the general protocol may need modification based on the antibody used and the subsequent assay that employs the immunodepleted extract. The following protocol is for rabbit polyclonal antibodies that bind avidly to Protein A agarose beads.

1. Wash Protein-A Sepharose CL-4B beads at least three times with large volumes of 1× PBS (beads to buffer ratio is about 1:10) to remove the sodium-azide present in the storage buffer. After each wash, the beads are pelleted by spinning at 800 × g for 1 min on a tabletop centrifuge.

2. Add 1–4 volumes of sera to one volume of beads. If affinity-purified antibodies are used, the volume of antibody added can be lower. The bead-antibody suspension is placed in spin columns and mixed thoroughly by incubating overnight at 4 °C on a rotator.

3. Wash the bead-antibody mixture in the same buffer that was used to make the extract. If the beads are washed in the spin column itself, then at least five washes are recommended. After each wash, the columns are placed in Eppendorf tubes and spun at 800 × g for 1 min to remove residual liquid. Care should be taken to prevent drying of the beads during these spins. If necessary, the time/speed of the spin should be reduced.

4. Resuspend the beads with the extract that needs to be depleted. Typically, for one volume of beads, 3–5 volumes of the extract are used, but this can vary depending on the protein being depleted. The column containing the bead/extract mixture is incubated at 4 °C on a rotator for 30–40 min.

5. To collect the extract, place the column in a fresh Eppendorf tube and spin at 800 × g for 1 min. The beads are retained in the column and the extract flows through into the Eppendorf tube. For additional rounds of depletion, the depleted extract is incubated in a new column containing a second batch of (antibody-treated) beads and **steps 4** and **5** are repeated. Complete depletion of most proteins requires two to three rounds of depletion.

6. The depleted extract must be compared to a control extract by Western Blotting to monitor the efficiency of depletion. The control extract is one which is subjected to the same procedure with control IgG instead of antibodies. Testing protein depletion levels are strongly recommended before using the depleted extract in assays designed to investigate protein function.

Alternatively, Protein A nonaggregating iron oxide magnetic beads can also be used, allowing for the depletion of small quantities of extract (≤20 μL) with minimal loss of volume through magnetic separation.

3.2 Preparation of DNA Templates

To study DNA replication and the DDR, *Xenopus* extracts are used in conjunction with a variety of DNA templates, including sperm chromatin and plasmid DNA. In this section, we describe the preparation of both chromosomal as well as small DNA templates typically used to investigate the DDR. Whereas intact templates are widely used to understand normal DNA replication, the DNA is sometimes subjected to damage in order to investigate damage-related physiological processes such as checkpoint and repair assays. Therefore, the preparation of damaged DNA templates is also discussed.

Chromatin templates are used in several assays to monitor DNA replication and the binding of DNA-associated proteins. In some instances, the DNA is appropriately damaged to investigate DNA damage-related phenomena. The protocols to make both intact and damaged chromatin are described below (Subheadings 3.2.1 and 3.2.2). Moreover, defined human chromosome segments using purified bacterial artificial chromosomes [39] have also been used in LSS extracts. This opens the possibility of performing biochemical and proteomics analyses using specific chromatin contexts.

The use of small DNA templates has specific advantages, notably the possibility of engineering site-specific alterations on the DNA that can be easily accomplished using synthetic oligonucleotides and plasmids. Furthermore, the introduced lesions can be designed to mimic the damage caused by UV radiation or chemotherapeutic agents such as mitomycin C and cisplatin. These templates, in combination with *Xenopus* extracts, notably the HSS/NPE system, proved to be extremely useful to study specific checkpoint signaling pathways and the repair of protein–DNA adducts [11].

The protocols to obtain two types of small DNA molecules are described in Subheadings 3.2.3 and 3.2.4 below.

3.2.1 Demembranated Sperm Chromatin

Demembranated sperm chromatin described below are used to study chromosome replication and as starting material to prepare NPE, a nucleus-free system used in replication, repair and checkpoint assays.

1. Prime male frogs with 25U PMSG (per frog) 3–5 days before use and inject with 125U HCG (per frog) the day before Sperm Chromatin Preparation (*see* **Note 1**). It is suggested that 5–6 male frogs be used to obtain adequate sperm of high concentration (200,000 nuclei/μL).

2. Put frogs in 0.05% benzocaine in water for 30 min.

3. Dissect frogs (incision on the ventral side), cut out testes and immediately place in cold 1× MMR in a petri dish on ice.

4. With a clean razor blade, remove fat tissue and blood vessels and rinse three times in MMR. The final testes should be as free from fat and blood tissues as possible. If necessary, use a dissecting microscope to dissect out unwanted tissue.

5. Wash three times in 1× NPB.

6. Remove buffer and macerate extensively with a clean, sharp razor blade. Thorough maceration is critical to obtain maximum recovery. Place the testes in a 60-mm plastic petri dish on ice and macerate for about 20 min.

7. Add 10 mL cold NPB and aspirate a few times with a plastic Pasteur pipette. Filter the sperm suspension through two layers of cheesecloth into a 50-mL Falcon tube. Squeeze excess liquid and any remaining sperm from the cheesecloth. Wash cheesecloth in 2–4 mL NPB and squeeze out any remaining sperm into the tube. Split the contents of the 50-mL tube into two 14-mL Falcon tubes (Becton Dickinson).

8. Centrifuge using an HB-6 rotor at 6,000 rpm (~6,000 × g) for 15 min at 4 °C to recover sperm and remove debris.

9. Remove supernatant and wash pellet with 10 mL NPB, trying to remove any blood vessels and debris. Repeat the wash process 2–3 times and spin for 10 min between washes. If blood vessels or other debris are still present, centrifuge at low speed (800 rpm) and transfer the liquid containing the sperm to a new tube. The debris should remain at the bottom of the tube. Recentrifuge at 6,000 rpm for 15 min.

10. Remove supernatant from the last wash and resuspend pellet in 1 mL NPB + 0.2% Triton X-100. Incubate for 15 min at room temperature with gentle mixing.

11. Add 10 mL of cold NPB + 3% BSA and mix gently to obtain a homogenous suspension. Centrifuge for 10 min at 6,000 rpm.

12. Wash two times with 10 mL NPB + 0.3% BSA.

13. Resuspend in 0.5 mL NPB + 0.3% BSA and 30% glycerol.

14. Take a 2 µL aliquot of sperm. Mix with 2 µL of DAPI fix and then dilute in 96 µL. Calculate sperm concentration with a hemocytometer. Sperm tend to sediment over time and sometimes to clump together, leading to inaccurate counts. Tap the tube before mixing or aliquoting but avoid vertexing or hard pipetting. Adjust sperm concentration to 200,000/µL with NPB + 0.3% BSA 30% glycerol.

15. Freeze nuclei using liquid nitrogen and store at −80 °C in aliquots (*see* **Note 14**).

3.2.2 Generation of Double-Strand Breaks (DSBs) in Chromatin

Xenopus extracts can be used to assess the physiological consequences of DNA damage and identify the protein complexes that assemble at the site of DNA damage. The use of S-phase and M-phase extracts (Subheading 3.1) further enables the identification of cell cycle "phase-dependent" composition of the DNA repair complexes. To introduce DNA damage in chromatin, we use PflMI restriction endonuclease [40, 41], although other suitable enzymes can also be used.

1. To the extract, add demembranated sperm nuclei/chromatin (2500–5000 nuclei/μL). Incubate for 10 min at 21 °C.

2. Remove 15 μL of extract into a separate Eppendorf tube. This is the "No-Damage" control.

3. To the remaining extract–sperm mixture, add 0.05 U/μL of PflMI restriction endonuclease. Incubate at 21 °C.

4. At desired time intervals, remove 15 μL of extract and stop the reaction by mixing with 800 μL of ice-cold chromatin isolation buffer (supplemented with 0.125% Triton X-100). Keep on ice for 5 min.

This protocol is used to perform both the chromatin binding assay to detect proteins involved in various stages of the cell cycle and in other checkpoint assays. HSS can also be used for these types of studies. Alternatively, DSBs can be generated using the radiomimetic drug neocarzinostatin (200 ng/μL), which generates breaks with heterogenous ends or etoposide (20 μM), which stabilizes Topoisomerase 2 covalent protein complexes at the DNA ends.

3.2.3 Preparation of Plasmids containing Interstrand Cross-Links (ICLs)

DNA interstrand cross-links (ICLs) can be induced by chemotherapeutic, environmental or endogenous agents. They represent a class of damaged DNA where the strands are covalently linked, usually leading to a distorted DNA helix. Interstrand cross-links are highly toxic since they can block DNA replication and transcription. Their removal or bypass is accomplished by the concerted effort of multiple repair pathways and repair failure can result in mutations, rearrangements, tumors, or cell death.

Checkpoint signaling induced by ICLs and their repair mechanisms have been intensively studied over the last decade using *Xenopus* cell-free extracts. The advantages of the *Xenopus* egg extract system mentioned above have been key for unraveling the molecular details that repair this class of genotoxic lesions.

Interstrand crosslinks can be generated on plasmid DNA with SJG-136 (NCI/Spriogen LTD), a rationally designed crosslinking chemical, that creates ICLs between opposing guanine residues [42].

1. Incubate pBS vector with 3 μM SJG-136 (NCI/Spriogen LTD) in 50 mM triethanolamine and 2 mM EDTA, overnight at 37 °C.

2. Purify DNA by ethanol precipitation and resuspended in water.

3. Assess the quality and quantity of plasmids recovered by agarose gel electrophoresis with ethidium bromide.

Alternatively, is it possible to obtain a plasmid with a single interstrand crosslink by incubating specific oligonucleotides with a single pair of opposing guanines with SJG-136 and then cloning the resulting crosslinked duplex product onto a plasmid backbone [42].

3.2.4 Preparation of DNA with Double-Strand Breaks

DNA molecules with DSBs have been generated in our laboratory by using either restriction enzymes or PCR. To obtain DNA fragments with DSBs, we use the circular pBR322 plasmid as a template and digest it to completion with restriction endonucleases. We tested different enzymes generating different types of DNA ends (blunt, 3′ overhang, or 5′ overhang) and did not observe differences in the DNA-damage response [43]. Such DNA can be used in assays that investigate the effect of damage on DNA replication and also in chromatin binding experiments.

1. Digest 0.5 mg of pBR322 with HaeIII (NEB). HaeIII cuts pBR322 plasmid 25 times, thus generating 26 fragments containing 2 DSBs each.

2. Digest DNA and extract twice in phenol–chloroform, then precipitate in ethanol and sodium acetate.

3. Resuspend DBS-containing DNA in water at a concentration of 1 mg/mL.

4. Dilute the DSBs stock solution into the extracts to the desired concentration.

Alternatively, we have used λ-DNA that was digested with a series of restriction enzymes giving rise to different numbers of restriction fragments. λ-DNA is digested with XbeI, NcoI, HindIII, and BstEI enzymes that generate 2, 5, 7, and 14 fragments, respectively. This approach enables us to increase the concentration of DSBs in the extracts while keeping the mass of added DNA constant. To obtain 1 kb DNA fragments by PCR, we use the M13 ssDNA template using 22 nt primers complementary to positions 5570 and 6584, as described before [44].

3.2.5 Preparation of Biotinylated Substrates

Biotinylated substrates are extensively used in replication, checkpoint and repair assays. We have established a system in which we use biotinylated DNA substrates to understand the initiation of DNA replication and DNA damage checkpoint signaling [45].

1. One microgram of the gel-purified DNA fragment is end-labeled with 1-unit T4 polymerase in the presence of 33 μM each of dATP, dGTP, dTTP, and biotin–dCTP for 15 min at 12 °C.

2. Stop reaction by addition of 50 mM EDTA and incubate at 76 °C.

3. Purify the labeled DNA using the PCR purification kit as per the manufacturer's instructions.

4. Quantify the purified DNA by UV absorbance at 260 nm.

In our experience, we observed that DNA damage response and checkpoint activation were induced by the relatively short substrates [45], thus increasing the potential use of such substrates in checkpoint studies.

3.3 Use of the Xenopus System to Study Cell Cycle Checkpoints/Checkpoint Assays

The maintenance of genome integrity in cells is constantly challenged by DNA damaging agents. The response to such DNA lesions involves the prompt detection, signaling and repair of the damaged DNA. Two proteins with established roles in damage signaling are the ATM and ATR protein kinases [6].

The ATM- and ATR-dependent checkpoints can be classified based on whether or not they depend on active DNA replication and fork progression. During G1 or S phases, double-strand breaks (DSB) and ssDNA can induce replication-independent checkpoints, namely the G1/S checkpoint and the Intra-S checkpoint. However, there also exist replication-dependent checkpoints such as the Replication checkpoint and the S/M checkpoint [46].

Replication-independent checkpoints: The G1/S checkpoint can result from IR or radiomimetic agents and functions to prevent origin firing until the damage is repaired. This pathway can be either p53 dependent or independent. The two p53-independent pathways involve ATM and ATR: Firstly, the presence of DSBs activates an ATM-dependent checkpoint resulting in the phosphorylation and inhibition of Cdk2 activity. This inhibition of Cdk2 prevents the loading of Cdc45 onto chromatin and blocks origin firing [46]. The second p53-independent G1/S checkpoint ensues upon the generation of aberrant DNA structures comprising of ssDNA-RPA intermediates generated in G1. Such intermediates can be generated experimentally by treatment with exonuclease III or by addition of etoposide, an inhibitor of DNA topoisomerase II [41, 47]. This RPA-dependent signaling results in ATR activation, which in turn inhibits origin firing [48].

The Intra-S checkpoint inhibits late origin firing. The proper installation of this checkpoint requires the phosphorylation of serine residues 278 and 343 of Nbs1 by ATM. After sensing the damage, ATM and ATR activation are followed by phosphorylation of their targets- Chk2 and Chk1, respectively. The ultimate result of the ensuing cascade is to downregulate Cdk2 and Cdc7 protein kinases, which in turn prevent cdc45 loading and origin activation.

S-phase checkpoint: The S-phase checkpoint is activated in response to UV, IR and HU treatments and usually requires that active DNA replication be elicited. This checkpoint results in ATR activation and phosphorylation of its target Chk1 on serine residues 317 and 345 [49]. The phosphorylation of Chk1, in turn, culminates in the phosphorylation-dependent degradation of Cdc25A and inhibition of Cdk2-Cyclin E [50]. Other regulators of the S-phase checkpoints are Claspin/Mrc1, MRN complex, BRCA1 and FANC proteins. The phosphorylation of Nbs1 by ATM is another event crucial for the S-phase checkpoint [51].

Replication-dependent checkpoints: The Replication checkpoint and the S/M checkpoint are other replication-dependent checkpoints that are active during the S phase. The Replication checkpoint is initiated by stalled forks in response to genotoxic stresses, aberrant DNA structures or DNA damage and inhibits origins through the activity of ATR and Chk1. The S/M checkpoint ensures that DNA replication is complete before the cell enters mitosis. This checkpoint prevents mitotic entry by enabling Chk2 phosphorylation and activation, which in turn inhibits Cdc25 phosphatase activity and prevents the activation of the Cyclin B-cdk1 complexes. The S/M checkpoint also involves ATM substrates such as Nbs1 and BRCA1 [52, 53].

Replication stress, which is defined as the slowing and stalling of replication forks during DNA replication, is a potent inducer of ATR mediated DNA damage response. Defects in ATR pathway signaling are associated with replication fork collapse resulting in double-strand breaks that can lead to chromosomal rearrangements and genome instability [54]. Replication stress is induced in cellular systems with hydroxyurea, a ribonucleotide reductase inhibitor that quickly leads to dNTP pools exhaustion. Since the egg stockpile of dNTP pools in the extracts is remarkably high, hydroxyurea is ineffective in fork slowdown in extracts. However, aphidicolin, an inhibitor of replicative polymerases, is a potent activator of ATR in extracts replicating chromatin templates. Additionally, fork collapse can be mimicked in the extracts by treatment with nicking endonucleases such as S1 or mung bean nucleases that result in one-ended double-strand breaks after fork passage [55].

G2/M checkpoint: The G2/M checkpoint targets Cdk1-Cyclin B kinase and prevents mitotic entry in response to DNA damage or incomplete S phase. Cell cycle arrest, in this case, involves both the ATM and the ATR pathways [56]. The G2 checkpoint is abrogated by caffeine and is sometimes used as an assay to test the involvement of the ATM pathway.

In addition to the above functions, the ATM/ATR kinases also assist in DNA repair by inducing repair proteins and activating them by post-translational modifications such as phosphorylation. The activation of ATM is, in turn, an indicator of checkpoint activation and can be assayed by assessing the phosphorylation of its substrates such as Smc1, Chk2, or H2AX.

The above paragraphs (in Subheading 3.3) are a few examples of the proteins involved in and the mechanisms that drive various cell cycle checkpoints activated as a result of DNA damage. To assess whether a checkpoint has been activated or to investigate checkpoint-related phenomena further, the proteins involved, their levels, phosphorylation status, and other modifications are frequently used as readouts. For instance, phosphorylation of Chk1, Chk2, or H2AX indicates that the ATR, ATM, or the ATM/ATR pathways have (respectively) been activated.

The *Xenopus* cell-free system can be used to investigate cell cycle checkpoints in a variety of ways such as monitoring DNA replication, studying the phosphorylation or activation of target proteins, and the assembly or localization of repair proteins on the site of DNA damage. The following sections describe the assays used to study checkpoints.

DNA replication assay to study checkpoint activation: Extracts derived from *Xenopus* eggs can support semiconservative DNA replication of genomic DNA upon the addition of DNA templates [17]. However, in the presence of DNA damage, checkpoints are activated. These checkpoints prevent the initiation or progression of DNA replication upon DNA damage [57, 58] and coordinate DNA replication, recombination and repair processes [6, 59]. This inhibition occurs both when the damage is present on the template during replication in "cis" as well as when DNA damage signaling is induced in "trans" by DNA containing exogenous DSBs. Both types of DNA damage can activate a checkpoint. The use of the *Xenopus* system to investigate such checkpoints is described below.

3.3.1 DNA Damage Response Induced by Single-Strand DNA Gaps in Chromatin

We have developed a cell-free system that recapitulates the inhibition of DNA replication in the presence of single-strand DNA gaps [60, 61]. Single-strand DNA gaps are generated by incubating chromatin in cell-free extracts in the presence of etoposide, an inhibitor of topoisomerase II, or by in vitro treatment of chromatin by DNA exonuclease III. Etoposide generates lesions in the chromatin templates that are undergoing DNA replication [47] by blocking the activity of DNA topoisomerase covalently linked to DNA 5′ termini [41, 62].

In addition to studying checkpoints induced by ssDNA gaps in chromatin, the protocol described below (the portion that describes the monitoring of DNA replication) can also be used to assess the replication of chromatin and/or small DNA templates in various extracts.

1. Incubate 20 μL of activated extract with 5000 demembranated sperm nuclei/μL in the presence of etoposide at 21 °C for 90 min (*see* **Note 15**).

2. Concentrations of etoposide ranging from 10 to 50 μM are effective in inducing a checkpoint response, as seen by the inhibition of genomic DNA replication.

3. Etoposide-induced inhibition of DNA replication is rescued by the addition of 5 mM caffeine, an inhibitor of checkpoint signaling kinases, including ATM and ATR [63].

4. Monitor DNA replication by incorporation of α-^{32}P-dCTP into the chromatin. Add 0.2 μCi of α-^{32}P-dCTP to each replication reaction.

5. Stop DNA replication reactions by diluting the samples in 200 μL of stop solution.

6. Incubate the samples with 1 mg/mL of proteinase K for 30 min at 37 °C.

7. Extract DNA with 1 volume of phenol–chloroform.
8. Centrifuge the samples for 10 min at room temperature.
9. Recover the aqueous phase and precipitate with 2 volumes of ethanol and 1/10 volume of sodium acetate (3 M stock).
10. Resuspend the pellet in DNA loading buffer and run on a 0.8% agarose gel in TBE.
11. Fix the gel in 7% TCA (some protocols require a higher percentage of TCA). Position the gel between two layers of Whatman 3 MM paper and stacks of filter paper. Dry the gel overnight on the bench.
12. Expose the dried gel for autoradiography.

3.3.2 DNA Damage Responses Induced by DSB in Chromatin

To investigate the cell cycle response to DNA damage at the onset of S phase, we use the *Xenopus* cell-free system designed to study the initiation of DNA replication [64]. Activated extracts are treated with either circular plasmid DNA, plasmid DNA containing DSBs, or λ-DNA containing DSBs. The treatment of the cytosolic extracts with DSBs-containing DNA activates a checkpoint in *trans*. In this protocol, the damaged DNA is only used to trigger the checkpoint and is subsequently removed, following which the extract is tested for its ability to replicate intact DNA templates. The damaged template is removed to avoid any interference with genomic DNA replication, such as titration of essential factors required in the elongation step of genomic DNA replication.

1. Incubate 100 μL of activated LSS extract at 21 °C in the presence of 50 ng/μL of circular plasmid DNA or digested plasmid (DSB) for 15 min to activate the checkpoint (*see* **Note 15**).
2. For rescue experiments, extracts are pretreated with 5 mM caffeine to inhibit PIKK kinases nonspecifically, 100 μM KU55933 to specifically inhibit ATM, or 50 μM VE821 to inhibit ATR specifically. Our lab has also been successful in inhibiting ATM activity by the addition of purified anti-ATM antibodies for 15 min at 21 °C, before incubation with the damaged DNA.
3. For replication assays, add 0.2 μL α-^{32}P-dCTP and sperm chromatin (2000/μL) to 10 μL of each extract treated with different types of DNA molecules, caffeine, or specific ATR/ATR inhibitors.
4. Incubate the reactions for 30 min at 21 °C.
5. Stop reaction by adding 200 μL of Stop Solution. Mix well.
6. Add 1/20 volumes (10 μL) of Proteinase K and incubate at 55 °C for 1–2 h.
7. Cool samples to room temperature and spin down for 10 s.

8. Add 200 μL of phenol–chloroform–isoamyl alcohol (25:24:1) and mix thoroughly by inverting several times.

9. Centrifuge samples at 13,000 rpm (Sorvall HB-6 rotor, ~28,000 × g) for 10 min. Transfer the aqueous phase to a fresh tube. Add 1/10 volumes (20 μL) of sodium acetate (3 M Stock) and 2.5 volumes (500 μL) of cold 100% ethanol. Incubate at −20 °C overnight.

10. Spin samples for 10 min at 13,000 rpm.

11. Discard supernatant, briefly dry pellet and dissolve it in 15-μL DNA loading buffer with bromophenol blue and cyan blue dyes.

12. Load samples and electrophorese them on a 0.8% agarose gel.

13. Treat gel with trichloroacetic acid (Sigma) 50% made up in water for 20–30 min.

14. Dry gel overnight between 3 M blotting paper and several layers of absorbent paper towels. The next day, the gel may be dried for an additional hour at 70 °C using a vacuum dryer.

15. Subject dried gel to autoradiography; multiple exposure times may be required, ranging from 15 min to overnight.

3.3.3 Assay to Study Checkpoint Induced by ICL

To investigate whether the damaged DNA activates an ATM/ATR checkpoint, replication assays are conducted in the presence of caffeine, which is an inhibitor of PIKK kinases (ATR, DNA-PK and ATM) or specific inhibitors. The experiment described below uses DNA containing an interstrand crosslink; however, the protocol can be modified to investigate whether other damaged DNA templates similarly activate the ATM/ATR checkpoints. The protocol for the ICL induced damage assay is described below:

1. *Xenopus* extracts (HSS and NPE) are prepared as described above (Subheadings 3.1.5 and 3.1.6). Experiments typically use a ratio of HSS to NPE at 1:2.

2. Take two sets of two tubes. To each set of tubes, add 3 μL of HSS, 0.1 μL of NPE–ATP regeneration system (Subheading 2.6), and 5 ng control DNA (tube 1) or 5 ng ICL DNA (tube 2) (Control and ICL DNA as described in Subheading 3.2.3). To the first set of tubes, 5 mM caffeine is also added. The second set serves as a "no-caffeine control," and only buffer is added instead. Mix. Incubate for 30 min at 21 °C.

3. Add 6 μL NPE and 0.2 μL α-^{32}P-dCTP, mix by pipetting and incubate for 2 h at 21 °C.

4. Monitor DNA synthesis by the incorporation of ^{32}P for 90 min at 21 °C, followed by agarose gel electrophoresis (as described in Subheading 3.3.1; **steps 4–12**).

3.3.4 Study of the Checkpoint Induced by DNA Damage In-Trans on DNA Replication

Checkpoint activation can be assessed by monitoring the effect of the ICL DNA (or other forms of damaged DNA) on the replication of an intact plasmid. The experiment described below uses ICL DNA; however, a similar protocol can be used to study the impact of other types of damaged DNA on the replication of intact DNA templates.

1. Prepare *Xenopus* extracts (HSS and NPE) as described (Subheadings 3.1.5 and 3.1.6).
2. Add 3 μL of HSS, 0.1 μL of NPE–ATP regeneration system (Subheading 2.6), and 5 ng Control or ICL DNA. Mix. Incubate for 30 min at 21 °C.
3. Add 6 μL NPE; mix by pipetting and incubate for 1 h at 21 °C (Try to keep HSS to NPE ratio to 1:2).
4. Add the plasmid that is to be replicated to each of the tubes and also add 0.2 μL α-^{32}P-dCTP. Mix well and incubate at 21 °C for 20 min.
5. Analyze the replication of the plasmid DNA by agarose gel electrophoresis as described earlier (Subheading 3.3.1).

Since the two plasmids (control/ICL plasmid and the plasmid to be replicated) are of different sizes, they can be distinguished by gel electrophoresis.

3.3.5 Study of Phosphorylation to Assess Checkpoint Activation

It is known that DNA damage blocks DNA polymerases, allows for DNA unwinding by MCM helicases, and triggers an ATM or ATR-dependent checkpoint [65, 66]. The ATR checkpoint pathway is activated during S phase to sense and coordinate cellular responses to DNA damage. Several checkpoint proteins are recruited to the site of damaged DNA and play a role in the assembly of the ATR kinase signaling complex [67, 68]. The activated ATR phosphorylates its target proteins, including Chk1, to initiate the cellular response to the DNA damage. This phosphorylation of Chk1 is commonly used as a readout for ATR activation. In *Xenopus*, Chk1 phosphorylation at the S344 site is observed in response to DNA damage induced by UV rays, methyl methanesulfonate (MMS), 4-nitroquinoline 1-oxide (4-NQO), interstrand crosslinks (ICL), and aphidicolin [19, 69, 70]. This is best detected by a phospho-specific antibody capable of detecting *Xenopus* Chk1 phosphorylated at S344.

The phosphorylation of Histone H2AX at the Serine 139 residue is another frequently used indicator of DNA damage. Histone H2AX is a well-characterized substrate for activated protein kinases and is phosphorylated by both ATM (in response to IR) and ATR kinases (in response to UV) or DNA replication blocks [71, 72] at Serine 139 [73, 74]. In combination with 53BP1, histone H2AX is also thought to play a role in G2-M checkpoint activation [75]. In our laboratory, we have monitored the phosphorylation of both endogenous histone H2AX as well as exogenous H2AX peptide [20, 76].

Several other proteins get phosphorylated upon DNA damage. For instance, MRE11, a protein that is involved in the repair of double-strand chromosome breaks [77] and Nbs1 that forms a complex with MRE11 and Rad50 are both phosphorylated in response to DNA damage [78]. The above examples collectively suggest that the phosphorylation of various proteins can be used to assess the presence of DNA damage as well as, in some instances, determine the activation of cell cycle checkpoints. The use of phosphorylation assays and *Xenopus* cell-free extracts in checkpoint studies are outlined in the following sections (Subheadings 3.3.5–3.3.8).

3.3.6 Phosphorylated Histone H2AX Detection (Endogenous Substrate)

The detection of phosphorylated Histone H2AX (γH2AX) is a sensitive assay to monitor DNA damage-induced checkpoint signaling.

1. Incubate 50 µL of interphase extract (or extract in which the occurrence of DNA damage will be assessed) with 10,000 nuclei/µL for 90 min at 21 °C.

2. Isolate post replicative chromatin by diluting the extracts in chromatin isolation buffer containing 1 mM NaF, 1 mM sodium vanadate, and 0.125% Triton X-100.

3. Layer samples onto sucrose cushions (chromatin isolation buffer containing 30% sucrose and lacking Triton X-100), then spin at 6000 × *g* for 20 min at 4 °C.

4. Prepare a positive control by incubating sperm nuclei for 30 min in an interphase extract to decondense chromatin.

5. Isolate the chromatin and digest for 4 h with *Not*I restriction endonuclease.

6. Reisolate the digested chromatin through a sucrose cushion and incubate in interphase extract for 60 min.

7. Boil chromatin in Laemmli buffer and process for SDS-PAGE electrophoresis.

8. Use anti-γH2AX antibody for Western blotting at 1/2000 dilution.

3.3.7 Histone H2AX Phosphorylation Assay (Exogenous Substrate)

Another assay to monitor checkpoint signaling is the measurement of H2AX peptide phosphorylation [76]. Histone H2AX is a well-characterized substrate for protein kinases that are activated by DNA damage and is phosphorylated in vivo at serine 139 by ATM and ATR. We have reported earlier the use of the C-terminal peptide of mouse H2AX as a reporter substrate to monitor the response to damage [43]. The actual assay involves incubation of interphase extracts (LSS/Interphase extracts I/II or Activated CSF) with fragmented DNA and either wild-type (AVGK KASQASQEY) or control/mutant (AVGKKAAQAAQEY) H2AX peptides. The presence of fragmented/damaged DNA results in rapid phosphorylation of the exogenous peptide.

1. The extract (LSS or Activated CSF extracts) is incubated with DNA fragments to elicit a DSB response. The DNA fragments we use are biotinylated at one end and immobilized on Streptavidin magnetic beads (Dynabeads, Thermo Fisher) at concentrations of 20 and 60 ng/µL (corresponds to 1.2×10^{11} and 3.6×10^{11} ends per microliter, which in turn simulate irradiation doses of 70 Gy and 210 Gy respectively for a human lymphocyte).

2. Separate the DNA-bound beads from the extracts (according to the manufacturer's instructions) and wash DNA six times with ELB buffer supplemented with 0.1% (v/v) Triton. The resulting DNA-bound and soluble fractions of the extracts can be evaluated for various parameters.

3. Mix 2 µL of the above extract with 20 µL of EB buffer supplemented with 50 µM ATP, 0.4 µL γ^{32}P-ATP (stock—10 mCi/mL), and 1 mg/mL of either wild-type or mutant H2AX peptide (Sigma).

4. Incubate samples for 15 min at 30 °C.

5. Stop reactions by adding 2 µL of 0.5 M EDTA.

6. Spot reactions on p81 phosphocellulose filter papers (Upstate Biotechnology).

7. Wash filters three times in 5% (v/v) acetic acid and twice with water.

8. Air dry filters and quantify radioactivity in a scintillation counter.

It should be noted that the c.p.m. incorporated into the control mutant H2AX peptide is subtracted from the c.p.m. of the wild-type H2AX peptide and the values are normalized to those from control extracts treated with streptavidin beads in the absence of DNA (no DSB). Second, the extracts may be depleted of a specific protein by using antibodies (*see* Subheading 3.1.7), and the depleted extract can be analyzed as above to assess the function of a particular protein. Third, the extract can be preincubated with 5 mM caffeine to serve as a control in the H2AX peptide phosphorylation assay.

An alternate protocol for H2AX peptide phosphorylation is as follows:

1. Take 2 μL of LSS Extract (that has been preincubated with undamaged or damaged DNA as above) and mix with 20 μL of EB Kinase Buffer supplemented with 0.5 mg/mL histone H2AX peptide (wild-type or mutant), 50 μM ATP and 1 μL of γ^{32}P–ATP (10 mCi/μL, greater than 3000 Ci/mM).

2. Incubate samples at 30 °C for 20 min.

3. Stop reaction by adding 20 μL of 50% acetic acid.

4. Spot samples on p81 phosphocellulose filter papers.

5. Wash filters three times with 10% (v/v) acetic acid and twice with water. Air dry filters.

6. Quantify radioactivity using a scintillation counter.

7. Subtract the number of c.p.m. incorporated into the control mutant H2AX peptide from the number incorporated into the wild-type H2AX peptide; normalize the values to those from control extracts treated with streptavidin beads in the absence of DNA (negative control).

3.3.8 Phosphorylation of ATM/ATR Target Proteins

Response to DNA damage involves the sensing of the damage, transduction of the signal and activating the response. Each component of the damage response is carried out by specific proteins. The ATM and ATR proteins are master regulators of the pathway. The levels and phosphorylation status of these proteins and their downstream targets are commonly used readouts to investigate the activation of a checkpoint. For instance, to distinguish between ATR and ATM activation, phosphorylation of Chk1 protein at S344 and phosphorylation of ATM at S1981 can be investigated as follows:

1. Prepare *Xenopus* extracts (HSS and NPE) as described earlier (Subheadings 3.1.5 and 3.1.6).

2. Take two sets of two tubes. To each set of tubes, add 3 μL of HSS, 0.1 NPE–ATP regeneration system, and 5 ng control DNA (tube 1) or damaged DNA (tube 2). To the first set of tubes, 5 mM caffeine (made up in buffer) is also added. The second set serves as a "no-caffeine control," and only buffer is added instead. Mix. Incubate for 30 min at 21 °C.

3. Add 6 μL NPE and incubate for 90 min at 21 °C.

4. Analyze the soluble extracts by Western blotting with phospho-Chk1 antibody (Cell Signaling) and phospho-ATM antibody (Rockland Immunochemicals, PA).

To differentiate between ICL and canonical DNA replication checkpoints, the above experiment can be carried out with purified recombinant geminin (50 ng/mL stock, added at 1/50 final concentration) and roscovitine (500 μM in DMSO) instead of caffeine. We have observed that ICL induced Chk1 phosphorylation is resistant to geminin/roscovitine treatment indicating the existence of a replication-independent pathway [19]. To evaluate the role of the Fanconi Anemia (FA) pathway in the ICL-induced checkpoint, the above experiment is carried out with curcumin—an inhibitor of the FA pathway (100 μM in ethanol; Sigma).

Binding assays to study the localization of damage-dependent proteins: The association or binding of proteins on to chromatin or small DNA templates is frequently used to understand the composition of complexes that bind DNA and to investigate the function of a specific protein. Furthermore, understanding the binding patterns of proteins onto intact or damaged DNA can provide insights into DNA damage, signaling, and repair. The protocols to identify the proteins bound to either intact or damaged DNA (either chromatin or plasmid DNA) are described below.

3.3.9 Chromatin Binding Assay

A critical aspect of the DNA-damage response is the damage-dependent localization of proteins to the DNA. This is exemplified by the formation of damage-induced foci within the nuclei of mammalian cells. Cell-free systems allow the rapid mixing and subsequent separation of chromatin, nuclear, and cytoplasmic fractions. In chromatin binding assays, depleted extracts allow to investigate the function of a specific protein in coordinating the binding specificities of other proteins. To analyze the status of the proteins that are recruited to the chromatin in replication or checkpoint-dependent manner, chromatin binding assays can be performed.

1. Thaw the appropriate amount of LSS or HSS extracts.
2. Assemble chromatin in 15–50 μL of interphase extracts in which a checkpoint has or has not been activated.
3. Incubate 2500–5000 nuclei/μL for 30–120 min.
4. Following incubation, dilute each reaction in 800 μL of chromatin isolation buffer supplemented with 0.125% Triton X-100.
5. Lay diluted extracts over 350 μL sucrose cushions (chromatin isolation buffer containing 30% sucrose, omit the Triton X-100 in this sucrose-containing buffer).
6. Centrifuge the chromatin at 6000 × g for 20 min at 4 °C.
7. Carefully aspirate supernatant leaving a few μL at the bottom of the tube.
8. Add 0.5 μL (50 U) micrococcal nuclease and incubate min at RT. Avoid pipetting on the sidewalls of the tube. Transfer samples to clean tubes.
9. Add 2× Laemmli sample loading buffer and run the samples on SDS-PAGE.
10. Analyze by Western blotting with specific antibodies.

Either intact or damaged chromosomal DNA can serve as templates in the above assay. Optimal sperm concentration should be optimized for each experiment. In chromatin binding assays, there is the possibility of contaminating the chromatin-bound proteins with cytosolic proteins. Therefore, it required to include a "no DNA" control (extract processed exactly as other samples except that the DNA template or sperm nuclei are omitted from the reaction tube). The same protocol can also be conducted with HSS. For HSS/NPE, use a 1:2 ratio and demembranated sperm chromatin as the DNA template on which proteins are bound.

3.3.10 Binding Assay Using HSS/NPE and ICL Plasmid

To investigate the role of specific proteins in checkpoint signaling, we deplete the specific protein and determine the activation of downstream pathways and the phosphorylation of target proteins. Although the assay described below involves ICL plasmid, it must be noted that the protocol can easily be adapted to use chromatin or other small DNA templates instead of the ICL plasmid.

This assay is extensively applied to study the levels, phosphorylation status and identity of the proteins that are bound to DNA under intact or damaged conditions. We have used this assay in our laboratory to investigate checkpoint signaling from and the repair of a single, site-specific interstrand crosslink [19, 24, 42]. Alternatively, this assay can also be modified to carry out rescue experiments where recombinant wild-type or mutant protein is added back to the extract.

1. *Xenopus* HSS, NPE extracts prepared as described above ([79]).
2. Take 3 µL of HSS, 0.1 µL of NPE–ATP regeneration system, and 5 ng plasmid DNA (control DNA or the ICL DNA) in an Eppendorf tube. Mix. Incubate for 30 min at 21 °C. Add 6 µL NPE, mix well, and incubate for 1 h at 21 °C.
3. Dilute the reaction with 800 µL of Chromatin isolation buffer supplemented with Triton X 100.
4. Layer the entire mixture on 350 µL of 30% sucrose cushions made with chromatin isolation buffer (without Triton X-100). Low retention 1.5-mL Eppendorf tubes are used for this step.
5. Centrifuge at 6,000 rpm (6,000 × g. Use 8,000 rpm/10,000 × g for plasmid DNA templates) in HB-6 rotor for 30 min at 4 °C.
6. Carefully remove most of the liquid from the tubes after the spin but leave about 10 µL in the tube. This should contain the DNA/chromatin-bound protein. Freeze the tube in liquid nitrogen and store at −20 °C overnight.
7. Thaw out tubes with DNA bound protein and resuspend the pellets with 10 µL of 2× Laemmli sample buffer. Denature samples by boiling for 3 min and analyze by SDS-PAGE and Western Blotting.

A control sample without DNA must also be analyzed simultaneously to ascertain that there is no contamination from non–chromatin-bound proteins. A mock-depleted extract is also included as a control in these assays.

4 Notes

1. Depending on age and size, a female *Xenopus* will lay close to 50 mL of eggs, which results in between 5 and 10 mL of fully dejellied eggs. This yields between 2 and 4 mL of egg cytosol. Frogs are generally not fed after this initial priming in order to reduce the possibility of fecal matter contaminating the eggs. However, the experiments must be scheduled so that the initial priming and egg collection are done within a week so that the frogs are not deprived of eating for extended periods. Always follow your Institutional Animal Care and Use Committee guidelines.

2. Healthy eggs have a dark and a light hemisphere, almost symmetrical, with even and dense coloration (Fig. 1a). Throughout extract preparation, pale gray eggs, with no sharp hemispheres, floating or broken eggs must be discarded using a transfer pipette. Depending on age and size, a female *Xenopus* will lay close to 50 mL of eggs, which results in between 5 and 10 mL of fully dejellied eggs (Fig. 1b). This yields between 2 and 4 mL of clarified egg cytosol.

3. Gently swirl the eggs with a plastic pipette to facilitate the dejellying. Once completed, the total volume of eggs will be reduced by more than half (Fig. 1b), and you might notice transparent shells floating in the cysteine solution. Eggs are extremely fragile once the protective jelly coats are removed, handle with care. Over-dejellying of eggs while preparing any of the extracts is detrimental and will cause eggs to burst/pop and render them unusable. Typically, dejellying takes about 5–8 min. However, each batch of eggs is unique and therefore, dejellying times must be adjusted accordingly. Proper pH adjustment of the cysteine solution is critical, as slightly acidic solutions will not work. Quickly remove and discard any floating or white egg/shells with a plastic Pasteur pipette.

4. If Falcon tubes and HB-6 rotors are used, then the caps of the tubes should be removed before centrifugation to allow proper closing of the rotor's lid. We routinely prepare our extracts at 4 °C. Other groups have reported successful extract preparation at room temperature.

5. The cytoplasmic layer can also be pipetted by directly sliding a pipette tip against the wall of the tube through the top lipid

layer. Alternatively, the lipid layer can be removed first using a cotton swab.

6. Cytosolic extracts (CSF or activated) must be used immediately after preparation. Freezing and thawing the extract triggers apoptosis.

7. While freezing the Interphase extracts II, to avoid clumping of drops, it must be ascertained that the previous drop is completely frozen before adding the next drop. Frozen drops generally sink to the bottom of the P60 dish containing liquid nitrogen. The frozen drops can be picked with clean forceps and dropped into a frozen cryotube. Several drops can be placed in a single cryogenic vial. Do not fully close the vials unless all the liquid nitrogen in it has evaporated to avoid pressure rupture. Loosely place the cap and gently shake before fully screwing.

8. For the preparation of HSS, the MMR buffer can be made as a 10× stock, stored at 4 °C and diluted to 1× MMR. It is recommended to check that the pH of the diluted MMR is 7.7–7.8 before use.

9. For the ultracentrifugation step, if thin tubes are used, the catalog number is Beckman 326819.

10. If the NPE is contaminated by insoluble material that appears cloudy or opaque, the extract is recentrifuged in the TL55 rotor with Teflon adapters for 15 min.

11. It is recommended that NPE is flash-frozen in liquid nitrogen and stored in −80 °C as small aliquots. Repeated freeze-thaw cycles can result in loss of activity and must be avoided.

12. Demembranated sperm chromatin is stored at −80 °C in small aliquots (10 μL/tube) for replication assays and in larger aliquots (100 μL/tube) for making NPE. We routinely test before use sperm preparations to avoid background DNA damage and poor replication dynamics. Discard leftover sperm after thawing.

13. Pipetting the extracts gently but thoroughly is critical to the success of all procedures to achieve a homogenous extract. The formation of aggregates or incomplete mixing would lead to variable or irreproducible experimental results. Always avoid *vortexing* the extracts or introducing bubbles when working with these extracts, as this would result in protein oxidation and denaturation.

14. Alternatively, supplement activated extract with ExoIII chromatin at the same concentration of nuclei. ExoIII chromatin preparation: Incubate 10^6 sperm nuclei with 100 U DNA exonuclease III (Roche) for 10 min at 37 °C (in 60 mM Tris–

HCl, pH 8.0, and 0.6 mM MgCl$_2$ buffer); stop the reaction by addition of 1 mL NPB buffer and.

15. We routinely use 100 μM KU55933 [80] for ATM inhibition, 50 μM VE-821 [81] for ATR inhibition, or 100 μM NU7441 [82] for DNAPKcs.

Acknowledgments

This work was supported by grants from the National Institutes of Health P01 CA174653 and R35 CA197606 to J.G. and R50 CA233182 to T.A.

References

1. Jackson SP, Bartek J (2009) The DNA-damage response in human biology and disease. Nature 461(7267):1071–1078
2. Lanz MC, Dibitetto D, Smolka MB (2019) DNA damage kinase signaling: checkpoint and repair at 30 years. EMBO J 38(18): e101801
3. Goldstein M, Kastan MB (2015) The DNA damage response: implications for tumor responses to radiation and chemotherapy. Annu Rev Med 66:129–143
4. Murray AW (1995) The genetics of cell cycle checkpoints. Curr Opin Genet Dev 5(1):5–11
5. Thompson SL, Bakhoum SF, Compton DA (2010) Mechanisms of chromosomal instability. Curr Biol 20(6):R285–R295
6. Zhou BB, Elledge SJ (2000) The DNA damage response: putting checkpoints in perspective. Nature 408(6811):433–439
7. Costanzo V, Gautier J (2004) Xenopus cell-free extracts to study DNA damage checkpoints. Methods Mol Biol 241:255–267
8. Blow JJ, Laskey RA (2016) Xenopus cell-free extracts and their contribution to the study of DNA replication and other complex biological processes. Int J Dev Biol 60(7-8-9):201–207
9. Sannino V, Pezzimenti F, Bertora S, Costanzo V (2017) Xenopus laevis as model system to study DNA damage response and replication fork stability. Methods Enzymol 591:211–232
10. Parker MW, Botchan MR, Berger JM (2017) Mechanisms and regulation of DNA replication initiation in eukaryotes. Crit Rev Biochem Mol Biol 52(2):107–144
11. Hoogenboom WS, Klein Douwel D, Knipscheer P (2017) Xenopus egg extract: a powerful tool to study genome maintenance mechanisms. Dev Biol 428(2):300–309
12. Sannino V, Kolinjivadi AM, Baldi G, Costanzo V (2016) Studying essential DNA metabolism proteins in Xenopus egg extract. Int J Dev Biol 60(7–9):221–227
13. Cupello S, Richardson C, Yan S (2016) Cell-free Xenopus egg extracts for studying DNA damage response pathways. Int J Dev Biol 60(7–9):229–236
14. Willis J, DeStephanis D, Patel Y, Gowda V, Yan S (2012) Study of the DNA damage checkpoint using Xenopus egg extracts. J Vis Exp 69:e4449
15. Hashimoto Y, Costanzo V (2011) Studying DNA replication fork stability in Xenopus egg extract. Methods Mol Biol 745:437–445
16. Lebofsky R, Takahashi T, Walter JC (2009) DNA replication in nucleus-free Xenopus egg extracts. Methods Mol Biol 521:229–252
17. Blow JJ, Laskey RA (1986) Initiation of DNA replication in nuclei and purified DNA by a cell-free extract of Xenopus eggs. Cell 47 (4):577–587
18. Smythe C, Newport JW (1991) Systems for the study of nuclear assembly, DNA replication, and nuclear breakdown in Xenopus laevis egg extracts. Methods Cell Biol 35:449–468
19. Ben-Yehoyada M, Wang LC, Kozekov ID, Rizzo CJ, Gottesman ME, Gautier J (2009) Checkpoint signaling from a single DNA interstrand crosslink. Mol Cell 35(5):704–715
20. Costanzo V, Robertson K, Bibikova M, Kim E, Grieco D, Gottesman M, Carroll D, Gautier J (2001) Mre11 protein complex prevents double-strand break accumulation during chromosomal DNA replication. Mol Cell 8 (1):137–147
21. Walter J, Newport J (2000) Initiation of eukaryotic DNA replication: origin unwinding

22. Walter J, Sun L, Newport J (1998) Regulated chromosomal DNA replication in the absence of a nucleus. Mol Cell 1(4):519–529
23. Dominguez-Sola D, Ying CY, Grandori C, Ruggiero L, Chen B, Li M, Galloway DA, Gu W, Gautier J, Dalla-Favera R (2007) Non-transcriptional control of DNA replication by c-Myc. Nature 448(7152):445–451
24. Williams HL, Gottesman ME, Gautier J (2012) Replication-independent repair of DNA interstrand crosslinks. Mol Cell 47(1):140–147
25. Gautier J, Minshull J, Lohka M, Glotzer M, Hunt T, Maller JL (1990) Cyclin is a component of maturation-promoting factor from Xenopus. Cell 60(3):487–494
26. Gautier J, Norbury C, Lohka M, Nurse P, Maller J (1988) Purified maturation-promoting factor contains the product of a Xenopus homolog of the fission yeast cell cycle control gene cdc2+. Cell 54(3):433–439
27. Gautier J, Solomon MJ, Booher RN, Bazan JF, Kirschner MW (1991) cdc25 is a specific tyrosine phosphatase that directly activates p34cdc2. Cell 67(1):197–211
28. Lohka MJ, Masui Y (1983) Formation in vitro of sperm pronuclei and mitotic chromosomes induced by amphibian ooplasmic components. Science 220(4598):719–721
29. Murray AW, Kirschner MW (1989) Cyclin synthesis drives the early embryonic cell cycle. Nature 339(6222):275–280
30. Murray AW, Solomon MJ, Kirschner MW (1989) The role of cyclin synthesis and degradation in the control of maturation promoting factor activity. Nature 339(6222):280–286
31. Lupardus PJ, Van C, Cimprich KA (2007) Analyzing the ATR-mediated checkpoint using Xenopus egg extracts. Methods 41(2):222–231
32. Tutter AV, Walter JC (2006) Chromosomal DNA replication in a soluble cell-free system derived from Xenopus eggs. Methods Mol Biol 322:121–137
33. Maresca TJ, Heald R (2006) Methods for studying spindle assembly and chromosome condensation in Xenopus egg extracts. Methods Mol Biol 322:459–474
34. Field CM, Nguyen PA, Ishihara K, Groen AC, Mitchison TJ (2014) Xenopus egg cytoplasm with intact actin. Methods Enzymol 540:399–415
35. Costanzo V, Robertson K, Gautier J (2004) Xenopus cell-free extracts to study the DNA damage response. Methods Mol Biol 280:213–227
36. MacLean-Fletcher S, Pollard TD (1980) Mechanism of action of cytochalasin B on actin. Cell 20(2):329–341. https://doi.org/10.1016/0092-8674(80)90619-4
37. Trenz K, Errico A, Costanzo V (2008) Plx1 is required for chromosomal DNA replication under stressful conditions. EMBO J 27(6):876–885
38. Sparks J, Walter JC (2019) Extracts for Analysis of DNA Replication in a Nucleus-Free System. Cold Spring Harb Protoc 2019
39. Aze A, Sannino V, Soffientini P, Bachi A, Costanzo V (2016) Centromeric DNA replication reconstitution reveals DNA loops and ATR checkpoint suppression. Nat Cell Biol 18(6):684–691
40. Peterson SE, Li Y, Chait BT, Gottesman ME, Baer R, Gautier J (2011) Cdk1 uncouples CtIP-dependent resection and Rad51 filament formation during M-phase double-strand break repair. J Cell Biol 194(5):705–720
41. Aparicio T, Baer R, Gottesman M, Gautier J (2016) MRN, CtIP, and BRCA1 mediate repair of topoisomerase II-DNA adducts. J Cell Biol 212(4):399–408
42. Kato N, Kawasoe Y, Williams H, Coates E, Roy U, Shi Y, Beese LS, Scharer OD, Yan H, Gottesman ME, Takahashi TS, Gautier J (2017) Sensing and Processing of DNA Interstrand Crosslinks by the Mismatch Repair Pathway. Cell Rep 21(5):1375–1385
43. Costanzo V, Paull T, Gottesman M, Gautier J (2004) Mre11 assembles linear DNA fragments into DNA damage signaling complexes. PLoS Biol 2(5):E110
44. de Jager M, van Noort J, van Gent DC, Dekker C, Kanaar R, Wyman C (2001) Human Rad50/Mre11 is a flexible complex that can tether DNA ends. Mol Cell 8(5):1129–1135
45. Kurth I, Gautier J (2010) Origin-dependent initiation of DNA replication within telomeric sequences. Nucleic Acids Res 38(2):467–476
46. Kastan MB, Bartek J (2004) Cell-cycle checkpoints and cancer. Nature 432(7015):316–323
47. Costanzo V, Gautier J (2003) Single-strand DNA gaps trigger an ATR- and Cdc7-dependent checkpoint. Cell Cycle 2(1):17
48. Kim JM, Yamada M, Masai H (2003) Functions of mammalian Cdc7 kinase in initiation/monitoring of DNA replication and development. Mutat Res 532(1-2):29–40
49. Liu Q, Guntuku S, Cui XS, Matsuoka S, Cortez D, Tamai K, Luo G, Carattini-Rivera S, DeMayo F, Bradley A, Donehower LA, Elledge

SJ (2000) Chk1 is an essential kinase that is regulated by Atr and required for the G(2)/M DNA damage checkpoint. Genes Dev 14(12):1448–1459

50. Busino L, Chiesa M, Draetta GF, Donzelli M (2004) Cdc25A phosphatase: combinatorial phosphorylation, ubiquitylation and proteolysis. Oncogene 23(11):2050–2056
51. Zhao S, Weng YC, Yuan SS, Lin YT, Hsu HC, Lin SC, Gerbino E, Song MH, Zdzienicka MZ, Gatti RA, Shay JW, Ziv Y, Shiloh Y, Lee EY (2000) Functional link between ataxia-telangiectasia and Nijmegen breakage syndrome gene products. Nature 405(6785):473–477
52. Buscemi G, Savio C, Zannini L, Micciche F, Masnada D, Nakanishi M, Tauchi H, Komatsu K, Mizutani S, Khanna K, Chen P, Concannon P, Chessa L, Delia D (2001) Chk2 activation dependence on Nbs1 after DNA damage. Mol Cell Biol 21(15):5214–5222
53. Yarden RI, Pardo-Reoyo S, Sgagias M, Cowan KH, Brody LC (2002) BRCA1 regulates the G2/M checkpoint by activating Chk1 kinase upon DNA damage. Nat Genet 30(3):285–289
54. Zeman MK, Cimprich KA (2014) Causes and consequences of replication stress. Nat Cell Biol 16(1):2–9
55. Hashimoto Y, Puddu F, Costanzo V (2012) RAD51-and MRE11-dependent reassembly of uncoupled CMG helicase complex at collapsed replication forks. Nat Struct Mol Biol 19(1):17–U30
56. O'Connell MJ, Walworth NC, Carr AM (2000) The G2-phase DNA-damage checkpoint. Trends Cell Biol 10(7):296–303
57. Garner E, Costanzo V (2009) Studying the DNA damage response using in vitro model systems. DNA Repair (Amst) 8(9):1025–1037
58. Smith E, Costanzo V (2009) Responding to chromosomal breakage during M-phase: insights from a cell-free system. Cell Div 4:15
59. Cimprich KA (2003) Fragile sites: breaking up over a slowdown. Curr Biol 13(6):R231–R233
60. Costanzo V, Shechter D, Lupardus PJ, Cimprich KA, Gottesman M, Gautier J (2003) An ATR- and Cdc7-dependent DNA damage checkpoint that inhibits initiation of DNA replication. Mol Cell 11(1):203–213
61. MacDougall CA, Byun TS, Van C, Yee MC, Cimprich KA (2007) The structural determinants of checkpoint activation. Genes Dev 21(8):898–903
62. Burden DA, Osheroff N (1998) Mechanism of action of eukaryotic topoisomerase II and drugs targeted to the enzyme. Biochim Biophys Acta 1400(1-3):139–154
63. Sarkaria JN, Busby EC, Tibbetts RS, Roos P, Taya Y, Karnitz LM, Abraham RT (1999) Inhibition of ATM and ATR kinase activities by the radiosensitizing agent, caffeine. Cancer Res 59(17):4375–4382
64. Chong JP, Mahbubani HM, Khoo CY, Blow JJ (1995) Purification of an MCM-containing complex as a component of the DNA replication licensing system. Nature 375(6530):418–421
65. Pichierri P, Rosselli F (2004) Fanconi anemia proteins and the s phase checkpoint. Cell Cycle 3(6):698–700
66. Pichierri P, Rosselli F (2004) The DNA crosslink-induced S-phase checkpoint depends on ATR-CHK1 and ATR-NBS1-FANCD2 pathways. EMBO J 23(5):1178–1187
67. Melo J, Toczyski D (2002) A unified view of the DNA-damage checkpoint. Curr Opin Cell Biol 14(2):237–245
68. Cook JG (2009) Replication licensing and the DNA damage checkpoint. Front Biosci (Landmark Ed) 14:5013–5030
69. Kumagai A, Guo Z, Emami KH, Wang SX, Dunphy WG (1998) The Xenopus Chk1 protein kinase mediates a caffeine-sensitive pathway of checkpoint control in cell-free extracts. J Cell Biol 142(6):1559–1569
70. Lupardus PJ, Byun T, Yee MC, Hekmat-Nejad M, Cimprich KA (2002) A requirement for replication in activation of the ATR-dependent DNA damage checkpoint. Genes Dev 16(18):2327–2332
71. Fernandez-Capetillo O, Celeste A, Nussenzweig A (2003) Focusing on foci: H2AX and the recruitment of DNA-damage response factors. Cell Cycle 2(5):426–427
72. Paull TT, Rogakou EP, Yamazaki V, Kirchgessner CU, Gellert M, Bonner WM (2000) A critical role for histone H2AX in recruitment of repair factors to nuclear foci after DNA damage. Curr Biol 10(15):886–895
73. Rogakou EP, Pilch DR, Orr AH, Ivanova VS, Bonner WM (1998) DNA double-stranded breaks induce histone H2AX phosphorylation on serine 139. J Biol Chem 273(10):5858–5868
74. Burma S, Chen BP, Murphy M, Kurimasa A, Chen DJ (2001) ATM phosphorylates histone H2AX in response to DNA double-strand breaks. J Biol Chem 276(45):42462–42467
75. Fernandez-Capetillo O, Chen HT, Celeste A, Ward I, Romanienko PJ, Morales JC, Naka K, Xia Z, Camerini-Otero RD, Motoyama N,

Carpenter PB, Bonner WM, Chen J, Nussenzweig A (2002) DNA damage-induced G2-M checkpoint activation by histone H2AX and 53BP1. Nat Cell Biol 4(12):993–997

76. Dupre A, Boyer-Chatenet L, Gautier J (2006) Two-step activation of ATM by DNA and the Mre11-Rad50-Nbs1 complex. Nat Struct Mol Biol 13(5):451–457

77. Ogawa H, Johzuka K, Nakagawa T, Leem SH, Hagihara AH (1995) Functions of the yeast meiotic recombination genes, MRE11 and MRE2. Adv Biophys 31:67–76

78. Haber JE (1998) The many interfaces of Mre11. Cell 95(5):583–586

79. Shechter D, Costanzo V, Gautier J (2004) ATR and ATM regulate the timing of DNA replication origin firing. Nat Cell Biol 6(7):648–655

80. Hickson I, Zhao Y, Richardson CJ, Green SJ, Martin NM, Orr AI, Reaper PM, Jackson SP, Curtin NJ, Smith GC (2004) Identification and characterization of a novel and specific inhibitor of the ataxia-telangiectasia mutated kinase ATM. Cancer Res 64(24):9152–9159

81. Reaper PM, Griffiths MR, Long JM, Charrier JD, Maccormick S, Charlton PA, Golec JM, Pollard JR (2011) Selective killing of ATM- or p53-deficient cancer cells through inhibition of ATR. Nat Chem Biol 7(7):428–430

82. Cano C, Barbeau OR, Bailey C, Cockcroft XL, Curtin NJ, Duggan H, Frigerio M, Golding BT, Hardcastle IR, Hummersone MG, Knights C, Menear KA, Newell DR, Richardson CJ, Smith GC, Spittle B, Griffin RJ (2010) DNA-dependent protein kinase (DNA-PK) inhibitors. Synthesis and biological activity of quinolin-4-one and pyridopyrimidin-4-one surrogates for the chromen-4-one chemotype. J Med Chem 53(24):8498–8507

Chapter 9

Mammalian Cell Fusion Assays for the Study of Cell Cycle Progression by Functional Complementation

Jongkuen Lee and David Dominguez-Sola

Abstract

Cell cycle progression, or its arrest upon checkpoint activation, is directed by a complex array of cellular processes dependent on the diffusion of chemical signals. These signals regulate the onset of each cell cycle phase and prevent undesired phase transitions. Functional complementation is a robust strategy to identify such signals, by which mutant phenotypes are rescued through complementation with candidate factors. Here we describe a method that reclaims a five-decade old mammalian cell–cell fusion strategy of functional complementation to study the molecular control of cell cycle progression. The generation of cell–cell fusions (heterokaryons) allows for the analysis, via immunofluorescence, of cell cycle regulator dynamics and evaluating the effective rescue of cell cycle progression in specific genetic settings.

Key words Cell cycle progression, Cell–cell fusion, Heterokaryon, Functional complementation, Protein shuttling

1 Introduction

Cell cycle regulation relies on the distribution of biochemical signals between nucleus and cytoplasm, which in turn coordinate multiple cellular processes and cellular components critical for cell division [1–4]. This simple but key notion was first inferred from a series of elegant experiments reported by Rao and Johnson 50 years ago [5]. In their experiments, they fused mammalian cells at different stages of the cell cycle to then follow the behavior of each nucleus within fused cells. Fusion of S and G1 cells, for example, accelerated the initiation of DNA synthesis in G1 nuclei. Similarly, fusion of S and G2 cells showed that G2 nuclei were incapable of restarting DNA synthesis [5]. The results of these experiments were a first indication that the sequential and unidirectional architecture of cell cycle progression is determined by chemical signals diffusing between nucleus and cytoplasm.

In a way, the original experiments by Rao and Johnson could also be understood as functional complementation analyses

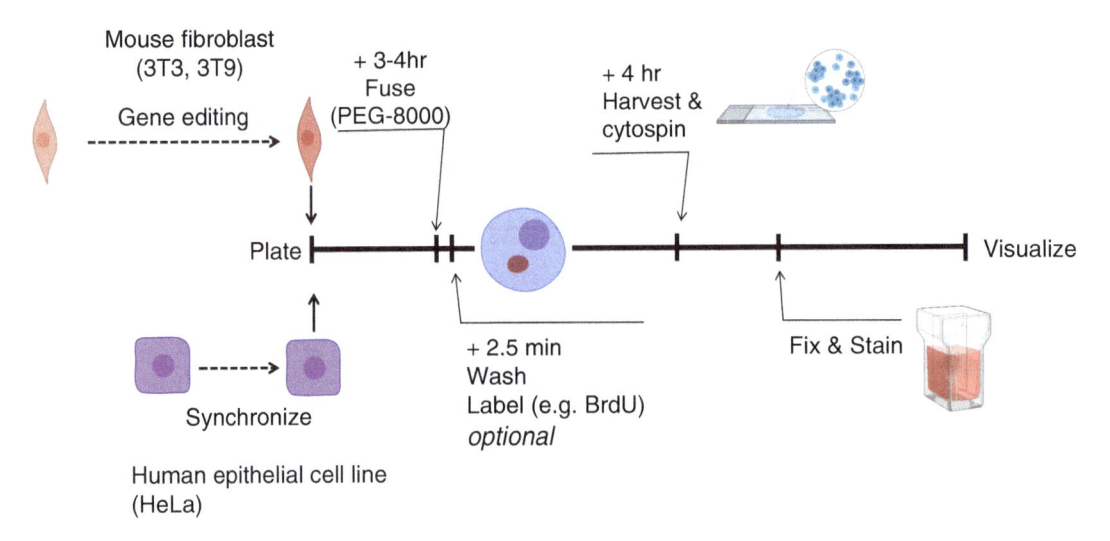

Fig. 1 Schematic representation of the experimental protocol

[6, 7]. Functional complementation implies restoring cellular functions or phenotypes by providing missing factors lost through genetic and/or epigenetic defects. Access to missing factors allows cells to accelerate processes or functions (e.g., triggering DNA synthesis in G1 nuclei) or typically, can rescue phenotypes in cells with specific functional deficits. Functional complementation has been widely used to study gene function and genotype–phenotype correlates. Historically, this strategy has been a key tool for the study of cell cycle and establishing the initial basis of the genetics of cell cycle control [8–10]. Classical functional complementation experiments use different methods of gene delivery to express gene products (synthetic or cloned from libraries generated from donor cells) into mutant cells (also known as "cloning by complementation") [6, 11]. Here we describe a method for functional complementation conceptually based on the original experiments of Rao and Johnson. This method uses cell fusion of donor and recipient cells to rescue cell cycle progression. A summary of the experimental strategy is shown in Fig. 1.

We originally used this strategy to investigate the role of MYC during the initiation of DNA replication (please refer to our original study) [12]. In this method we use immortalized mouse fibroblasts (recipient cells) and synchronized human HeLa cells (donor cells) to identify factors limiting S-phase entry, but it can easily be applied to other cell cycle transitions. Successful rescue of DNA synthesis in recipient (mouse) nuclei is visualized by incorporation of synthetic nucleosides. Shuttling of donor (human) proteins into mouse nuclei can also be confirmed through immunofluorescence. This specific phenomenon can also be visualized by expressing GFP-tagged proteins in HeLa donor cells via stable or transient transfection (Fig. 2). We envision that this protocol, with minor

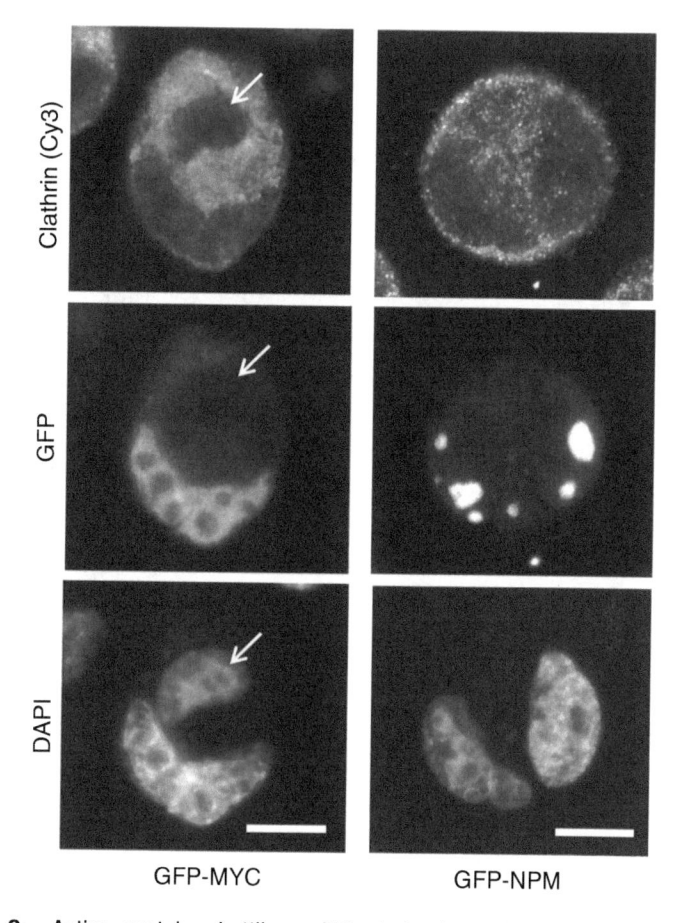

Fig. 2 Active protein shuttling within heterokaryons. H1299 cells were transiently transfected with GFP-tagged MYC or Nucleophosmin (NPM). 48 h post-transfection, we pretreated these cells with cycloheximide (CHX) for 1 h, and then fused to untransfected cells with PEG-8000, in presence of CHX (CHX treatment prevents new protein synthesis prior to or during fusion). The figure shows representative images of heterokaryons for each experiment. Note that unlike GFP-NPM (distributed in a nucleolar pattern), GFP-MYC (a fusion of a transcription factor) cannot travel to a neighbor nucleus within a single heterokaryon (arrow). NPM contains a nuclear export signal. Scale bar, 50 μm

modifications, can be used for unbiased discovery when combined with genetic manipulation of the donor and/or recipient cells. Cell fusion is a powerful method of cellular complementation, useful to study complex biological processes—for example, in the study of the molecular mechanisms required for cellular reprogramming [13].

2 Materials

2.1 Cell Lines

1. Genetically modified, immortalized mouse fibroblasts (*see* **Note 1**).
2. HeLa cells (ATCC CCL-2).

2.2 Medium, Buffers, and Supplements

1. Dulbecco's Modified Eagle's Medium with high glucose and L-glutamine (DMEM; different vendors), supplemented with 10% fetal bovine serum and penicillin/streptomycin.
2. Phosphate-buffered saline solution (PBS 1×, without calcium or magnesium).
3. 0.25% trypsin solution.
4. 50–52% polyethylene glycol (PEG) solution: Mix PEG-8000 (Sigma-Aldrich) at 50–52% weight/volume ratio in sterile PBS 1× and stir (warming up to 80 °C while stirring facilitates solubilization). Sterilize the solution by filtering through a 0.22-μm filter. Alternatively, a 50% sterile solution can be purchased from Sigma (Hybri-Max, Sigma-Aldrich Cat #7181).

2.3 Plastics and Equipment

1. Tissue culture dishes (for cell expansion and synchronization).
2. Multiwell cell culture plates (for cell–cell fusion).
3. Centrifuge polypropylene conical tubes (15 mL and/or 50 mL).

2.4 Chemical Reagents

1. Thymidine (Sigma-Aldrich). Prepare a 100 mM stock by dissolving 2.42 g of thymidine in 90 mL of sterile, tissue culture–grade PBS or H_2O. Adjust final volume to 100 mL and sterilize by filtration. Store aliquots. Use at 2 mM.
2. L-Mimosine (various vendors). Prepare a 10 mM stock solution in sterile PBS, sterilize by filtering and store at −20 °C. Use at 400 μM (*see* **Note 2**).
3. Nocodazole (Sigma-Aldrich). Prepare a 5 mg/mL stock solution in DMSO and store at −20 °C. Use at 40 ng/mL.
4. Cycloheximide (multiple vendors). Prepare a 10 mg/mL stock solution in DMSO or ethanol [14]. Sterilize by filtering and store at −20 °C. Alternatively, several vendors sell ready-made stock solutions. Use at 5–50 μg/mL (dose is adjusted to exposure time).
5. Polyethylene glycol 8000 (PEG-8000; multiple vendors). For details on preparation, see buffers section, above.
6. Optional: 5-bromo-2′-deoxyuridine (BrdU) or 5-ethynyl-2′-deoxyuridine (EdU). These two thymidine analogs can be used to detect active DNA synthesis and label cells in S-phase. Prepare a 10 mM stock solution in DMSO, and store at

−20 °C. They can be also dissolved in PBS at 10 mg/mL, sterilized by filtering and stored at −80 °C. Once thawed, the aqueous solution is stable for up to 4 months at 4 °C. Alternatively, 10 mg/mL stock vials are available for purchase from multiple vendors. Use at a final concentration of 10 μM.

2.5 Immunofluorescence

1. Fixative: 10% buffered formalin or 4% paraformaldehyde solution in PBS.
2. Wash buffer: 1× PBS with 0.1% Tween 20 (Sigma-Aldrich).
3. Blocking and permeabilization solution: 1× PBS plus 3% w/v bovine serum albumin (multiple vendors) plus 0.3% Triton X-100 (Sigma-Aldrich).
4. Anti-BrdU antibody (BD Bioscience or other vendors). An anti-BrdU antibody combined with DNase (BD Bioscience #340649) can be alternatively used.
5. Anti-Clathrin heavy chain antibody (multiple vendors—we have tested Cell Signaling Technology #2410, Santa Cruz Biotechnology #sc-12734).
6. Fluorochrome-labeled secondary antibodies (multiple vendors).
7. To stain nuclei/DNA: DAPI (4′,6-diamidino-2-phenylindole) 5 mg/mL solution (ThermoFisher Scientific). A 5 mg/mL stock corresponds to ~14.3 mM for DAPI hydrochloride and 10.9 mM for the dilactate. Store in cold, protected from light. Use DAPI at 3 μM for nuclei staining.

    Alternatively, nuclei can be stained using Hoechst 33342 solution (trihydrochloride, trihydrate, 1 mg/mL; ThermoFisher Scientific or BD Biosciences). A 20 mM (~2 μg/mL) aqueous solution is stable and can be purchased ready-to-use.
8. Aqueous slide mounting medium for fluorescence (e.g., Fluoromount—Millipore-Sigma; or Fluoromount-G—Electron Microscopy Sciences).
9. DNase I solution (1 mg/mL, BD Biosciences or Sigma-Aldrich. Dilute at 0.3 mg/mL in 1× PBS immediately prior to use).
10. Slide staining tray with humid chamber (e.g., StainTray, from multiple vendors).
11. Glass slides for microscopy (SuperFrost Plus, Fisher Scientific) and borosilicate glass coverslips (Fisher Scientific).
12. Cytospin funnels, white filter cards, and clips (Shandon/ThermoFisher Scientific).
13. Cytospin cytocentrifuge (ThermoFisher Scientific).

3 Methods

3.1 Cell Synchronization

3.1.1 Synchronization of HeLa Cells at the G1/S Transition by Double Thymidine/Mimosine Block

1. Make sure that cells are growing exponentially (50–70% confluency) at the time of synchronization. First, add thymidine solution to the growth medium to a final concentration of 2 mM, and incubate for 14–16 h (overnight).
2. Release block by washing the cells once in prewarmed 1× PBS and then in pre-warmed fresh medium, and then incubate in complete growth medium for 9 h (during the day).
3. After 9 h (full day), add 2 mM thymidine plus 400 µM L-mimosine and incubate for an additional 14–16 h (overnight). After this second block, >90% of cells are in G1/S (this can be confirmed by flow cytometry).
4. To collect cells in active S-phase, release cells for 1 h before harvesting. Alternatively, release cells for 8 h to obtain almost pure G2 cell pools (>90% by flow cytometry; *see* **Note 3**). Both thymidine and mimosine reversibly block cell cycle at the G1/S transition or right after cells start S-phase.

3.1.2 Synchronization of Cells in Mitosis or Early G1

1. Incubate cells in presence of 2 mM thymidine for 14–16 h (overnight).
2. Wash and release (as above) for 6–8 h in fresh medium; then directly add Nocodazole to a final concentration of 40 ng/mL and incubate for 12 h. This procedure arrests >70% of cells in early mitosis (prometaphase).
3. Pure mitotic cells can be collected by careful "shake-off" (hold the culture dish with the lid on, then hit on the sides with a flat hand a few times while you rotate the dish). Be gentle to avoid contamination with G2 cells.
4. Early G1 cells can be obtained from mitotic cells (after "shake-off") by replating these cells in fresh medium and allowing them to end mitosis for 3–5 h. Discard nonattached cells by gentle shake-off and wash with fresh medium.

3.2 Cell–Cell Fusion and Heterokaryon Formation

1. Prepare both cell types prior to the experiment so that they express the desired proteins (by transient transfection) or are synchronized at the desired cell cycle phase (*see* Subheading 3.1 step, above) (*see* **Note 1**).
2. Trypsinize cells and count. Briefly (a) discard growth medium; (b) wash once with 1× PBS solution and discard; (c) add a small volume of 0.25% Trypsin solution and (d); incubate at 37 °C (5% CO_2) for a few minutes until cells detach—the time is variable depending on the cell type used; (e) add complete growth medium to stop trypsin, resuspend thoroughly and proceed to count cells using a Neubauer chamber or cell counter (*see* **Note 4**).

3. In each well from a 6-well plate, plate a mixture 1:1 to 1:3 (recipient–donor) of cells to a total cell final number of 7×10^5 per well (see **Note 5**). Let cells attach for a few hours in the incubator (see **Note 6**).

4. Replace growth medium with fresh medium containing 50 μg/mL cycloheximide and incubate for 1 h. This will help discard nonattached cells and block new protein synthesis prior to fusion (see **Note 7**).

5. Wash cells with warm 1× PBS twice to eliminate residual medium (see **Note 8**).

6. Aspirate all PBS and add 1.5 mL of 50–52% PEG 8000 prepared in PBS and pre-warmed. Incubate at room temperature for 2:30 min to 3 min (see **Note 9**).

7. Carefully aspirate the PEG solution and wash thoroughly (three times) with warm PBS (see again **Note 10**), and twice (2×) with warm culture medium (see **Note 11**).

8. Replenish culture with fresh complete growth medium containing 50 μg/mL cycloheximide. Incubate for 3 h (see **Note 12**).

9. Optional step: Add BrdU to growth medium during the last 2 h of recovery—BrdU incorporation will label sites of new DNA synthesis (see **Note 13**).

10. Visualization and analysis (see **Note 14**). In our regular experiments, we trypsinize and cytospin all resulting cells. This strategy yields even cell spreads that facilitate the detection and analysis of heterokaryons under the microscope (see **Note 15**). Alternatively, fix and process coverslips.

3.3 Analysis of Heterokaryons by Immunofluorescence

We use immunofluorescence to assess BrdU incorporation in recipient nuclei and confirm and the successful shuttling of key cell cycle proteins within heterokaryons (i.e., Cyclin E) (Fig. 3). We refer to previously published protocols for additional details on immunofluorescence [15, 16].

Here below we describe a brief protocol to detect BrdU incorporation and highlight the cytosol of heterokaryons, which allows distinguishing fused from nonfused cells. Multiple combinations of specific antibodies can be adapted to this simple method.

1. Optional: Cytospin preparation. After trypsinization and resuspension in complete growth medium (DMEM+10%FCS), count and dilute cells at ~2×10^5/mL. Then transfer 200 μL per cytofunnel (regular size) or 500 μL (megafunnel). The larger the surface of cytospin, the larger the number of events. Spin at 800 rpm, 5 min. Disassemble cytospin sandwich and let slide dry for 30 s.

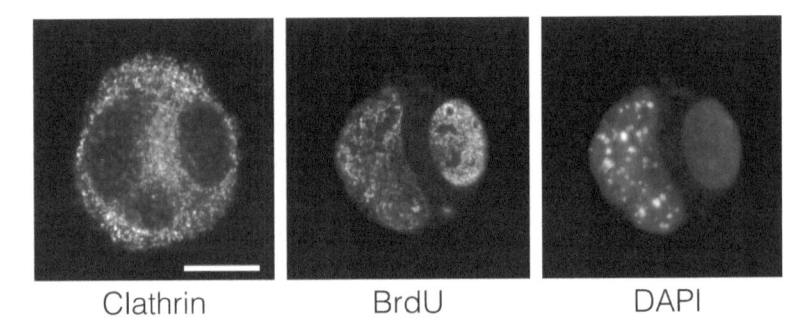

Fig. 3 Functional rescue of DNA synthesis in a mouse/human heterokaryon. We fused early G1 (post-mitosis) mouse 3T9 cells to cycloheximide-treated (human) HeLa cells arrested in early S-phase. After fusion, we labeled the cell culture with BrdU (10 μM) for 2 hours, in continued presence of cycloheximide. We then trypsinized, cytospun and fixed all cells for staining with antibodies against BrdU (DNA synthesis) and Clathrin (cytosol). Shown is one representative heterokaryon. The mouse nucleus is identifiable by a coarse speckled pattern revealed by the DAPI stain, in contrast to the fine chromatin of the HeLa nucleus. Incorporation of BrdU in the mouse nucleus confirms functional complementation (as observed originally by Rao & Johnson), which is attributable to factors contributed by the HeLa cell. Scale bar is 50 microns

2. Fix slides/coverslips postcytospin by immersing slides (20 min at room temperature).
3. Wash cells 3× with PBS–0.1% Tween, 5 min per wash.
4. Optional step: Denature DNA (*see* **Note 16**).
5. Block and permeabilize for 1 h in 3% BSA-0.3% Triton X-100-PBS (*see* **Note 17**).
6. Treat slides with DNase solution (0.3 mg/mL in 1× PBS), 1 h at 37 °C (*see* **Note 18**).
7. Wash slides once with 1× PBS–Tween. Remove excess liquid by taping edge on blot paper.
8. Incubate with prediluted primary antibodies for 2 h at room temperature or, alternatively, at 4 °C overnight (anti-BrdU, *see* **Note 18**; and anti-Clathrin, *see* **Note 19**).
9. Wash twice (2 × 5 min) with 1× PBS–Tween.
10. Incubate with fluorochrome-labeled secondary antibodies (diluted in 1× PBS–Tween) for 30–45 min at room temperature, in humid chamber.
11. Wash twice (2 × 5 min) in wash buffer.
12. Stain nuclei by incubating at room temperature with DNA dye (DAPI or Hoechst 33342) diluted in wash buffer (1× PBS–Tween) for 30 min at room temperature (*see* **Note 20**).
13. Rinse briefly in 1× PBS. Remove excess liquid by taping edge on blot paper.

14. Cover each slide with a glass coverslip and mount in aqueous medium for immunofluorescence and visualize in fluorescence or confocal microscope with appropriate filters.

4 Notes

1. Our original protocol was developed using HeLa cells and 3T3 or 3T9 mouse embryonic fibroblasts. HeLa cells are easy to synchronize by multiple methods, have rather large size and have been widely used to study cell cycle progression [17, 18]. However, other epithelial cell lines like H1299 cells (non-small cell lung cancer) are also suitable to fusion using this protocol (*see* Fig. 2). This may also be the case for other cancer cell lines or immortalized cells. The use of immortalized mouse fibroblasts has several advantages, including the easy distinction of mouse and human cell nuclei based on chromatin distribution using intercalating DNA stains. More importantly, the use of mouse fibroblasts makes feasible taking advantage of a large array of already available mouse genetic models—for example those generated in the Knock-out Mouse Project initiative—KOMP [19].

2. Alternatively, both thymidine and mimosine solutions can be prepared in DMEM and sterilized by filtering through 20-micron filters.

3. *Noteworthy trick*: addition of fresh medium plus serum stimulates S-phase entry.

4. Cells are trypsinized and mixed together to obtain an even mixture of cells and increase the chances to have cells from the two different types in close proximity.

5. Cell number should be carefully tested. In our experience, the number described above was quite efficient for the type of wells and plates used, but other densities might also work. The final goal is to have cells in relatively close proximity to facilitate membrane fusions. A ratio 1:3 was the most efficient in our experiments, but lower ratios may be also suitable for most assays (*see* below, **Note 14**).

6. This step can also be performed directly on pre-treated glass coverslips (e.g., poly-L-lysine coated), upon adjusting cell numbers based on the total surface covered.

7. The addition of cycloheximide inhibits new protein synthesis. This step ensures that all functional outcomes measured *after* fusion are driven by proteins present in the cells *before* fusion. This notion is important when we expect to detect evidence of nucleocytoplasmic shuttling of target proteins (*see* Fig. 2).

8. Alternatively, you can use serum-free medium at this step.

9. Mammalian cells can be fused by short incubations in 50% solutions of Polyethylene glycol (PEG 8000) in PBS. PEG 8000 is a highly hydrated polymer that can displace the water molecules between neighboring membranes. This effect facilitates exchange between lipid layers and eventual fusion of cell membranes when cells are in close proximity. The fine molecular mechanism by which PEG fuses cells is not well understood, though it is believed to cause small perturbations in lipids densities/packaging within contacting membrane leaflets. PEG molecules with other molecular weights can be used (50% solutions for 2–3 min). The choice of molecular weight and concentration should be empirically tested for each cell type [20]. In our experiments, PEG 8000 worked well for fusions between HeLa cells and mouse 3T9 fibroblasts [12]. Of note, multiple methods to promote cell fusion exist. The original experiments by Rao and Johnson [5] used Sendai virus to promote cell fusion, but methods alternative to Sendai virus of PEG-based fusion exist and may have some specific advantages [21]. The recent identification of dedicated proteins known as "fusogens," involved in physiologic cell fusion processes during development and tissue homeostasis, offers a larger repertoire of cell fusion strategies [22, 23].

10. Some cells will be unavoidably lost at this step.

11. Timing is critical in this step. In our experiments, shorter incubation times would fail to yield sufficient fusion events. Longer times can severely affect cell viability.

12. During this period cells will fuse, reorganize membranes and cytoskeleton, and heterokaryons will be formed. After 3 h some heterokaryons can be distinguished upon close inspection using phase contrast microscopy, appearing as multinucleated cells. This same procedure will generate homokaryons (fusions of two or more cells from the same type). A high number of homokaryons generally indicate that the ratio between the two cell types (**step 3**) is off and needs to be optimized.

13. Add BrdU or EdU to the culture medium at a final concentration of 3 μg/mL or 10 μM and incubate for 2 h. Incubation times are somewhat empirical, but in our hands 2–3 h can efficiently detect G1 nuclei as they transit to S phase upon complementation.

14. Fusion efficiency is usually low (we usually obtain 50 heterokaryons from a total starting number of one million cells). Since this assay is aimed visualizing changes on a cell-by-cell basis, 50 events provide substantial statistical power.

15. Human and mouse cell nuclei can be distinguished because of significant differences in chromatin configuration. Mouse cells

(particularly 3T3 or 3T9 fibroblasts) contain large heterochromatin regions that generate a pattern easily distinguishable using DAPI or Hoechst DNA stains (*see* Figs. 2 and 3). When fusing cells from the same species, it is necessary to label the nuclei of one of the cell lines with a specific marker. An efficient way to do this is to stably express H2B-GFP in one of the two cell lines being fused [24, 25].

16. DNA digestion or denaturation are necessary to facilitate the detection of incorporated BrdU. While our protocol uses an incubation step with DNase I, traditional protocols use DNA denaturation with hydrochloric acid (HCl) for the same purpose. Briefly, after fixation and washes, incubate slides in 2 N HCl (dilute from 12 N stock in dH$_2$O) and incubate at room temperature for 30 min. Rinse 3× over 10 min in 1× PBS, then proceed with blocking and permeabilization step. DNA digestion or denaturation is not required if you opt for EdU detection via click chemistry.

17. Although permeabilization is accomplished with a short incubation (5–10 min at room temperature) in 1× PBS plus 0.2–0.3% Triton X-100, we have observed more consistent staining when combining the permeabilization and blocking steps. *Additional note*: This is a good stopping point, if needed. Slides can be stored in 1× PBS in cold (4 °C) until the next day.

18. Do this in a humid chamber placed inside an incubator at 37 °C. Alternatively, this step can be combined with the anti-BrdU antibody incubation. If using EdU to label DNA synthesis, follow vendor guidelines for detection of EdU through the click reaction.

19. We use anti-Clathrin heavy chain antibodies to highlight cellular cytosols. This staining yields a granular cytosolic stain that spares nuclei. In our hands, this works well in HeLa cells and the resulting heterokaryons. Alternatively, other antibodies specific to cytosolic markers can be used.

20. Alternatively, DAPI or Hoechst dyes can be included during fluorochrome-labeled secondary antibody incubation (there is no need for a separate incubation).

Acknowledgements

Jongkuen Lee was supported by an Institutional Research Training Grant sponsored by the Tisch Cancer Institute at Mount Sinai. Research in our laboratory is funded by the National Cancer Institute and the National Institute of Allergy and Infectious Diseases (National Institutes of Health), the Gabrielle's Angel Foundation for Cancer Research, and the St. Baldrick's Foundation.

References

1. Barnum KJ, O'Connell MJ (2014) Cell cycle regulation by checkpoints. Methods Mol Biol 1170:29–40. https://doi.org/10.1007/978-1-4939-0888-2_2

2. Gookin S, Min M, Phadke H, Chung M, Moser J, Miller I, Carter D, Spencer SL (2017) A map of protein dynamics during cell-cycle progression and cell-cycle exit. PLoS Biol 15(9):e2003268. https://doi.org/10.1371/journal.pbio.2003268

3. Jackman, M, Kubota Y, den Elzen N, Hagting A, Pines J (2002) Cyclin A- and cyclin E-Cdk complexes shuttle between the nucleus and the cytoplasm. Mol Biol Cell 13(3):1030–1045. https://doi.org/10.1091/mbc.01-07-0361

4. Satyanarayana A, Kaldis P (2009) Mammalian cell-cycle regulation: several Cdks, numerous cyclins and diverse compensatory mechanisms. Oncogene 28(33):2925–2939. https://doi.org/10.1038/onc.2009.170

5. Rao PN, Johnson RT (1970) Mammalian cell fusion: studies on the regulation of DNA synthesis and mitosis. Nature 225(5228):159–164. https://doi.org/10.1038/225159a0

6. Nasmyth KA, Reed SI (1980) Isolation of genes by complementation in yeast: molecular cloning of a cell-cycle gene. Proc Natl Acad Sci U S A 77(4):2119–2123. https://doi.org/10.1073/pnas.77.4.2119

7. Reid BJ, Culotti JG, Nash RS, Pringle JR (2015) Forty-five years of cell-cycle genetics. Mol Biol Cell 26(24):4307–4312. https://doi.org/10.1091/mbc.E14-10-1484

8. Nasmyth K, Nurse P (1981) Cell division cycle mutants altered in DNA replication and mitosis in the fission yeast Schizosaccharomyces pombe. Mol Gen Genet 182(1):119–124. https://doi.org/10.1007/BF00422777

9. Nurse P, Hayles J (2019) Using genetics to understand biology. Heredity (Edinb) 123(1):4–13. https://doi.org/10.1038/s41437-019-0209-z

10. Nurse P, Thuriaux P, Nasmyth K (1976) Genetic control of the cell division cycle in the fission yeast Schizosaccharomyces pombe. Mol Gen Genet 146(2):167–178. https://doi.org/10.1007/BF00268085

11. Lundblad V (2001) Cloning yeast genes by complementation. Curr Protoc Mol Biol. Chapter 13:Unit13 18. https://doi.org/10.1002/0471142727.mb1308s05

12. Dominguez-Sola D, Ying CY, Grandori C, Ruggiero L, Chen B, Li M, Galloway DA, Gu W, Gautier J, Dalla-Favera R (2007) Non-transcriptional control of DNA replication by c-Myc. Nature 448(7152):445–451. https://doi.org/10.1038/nature05953

13. Tsubouchi T, Soza-Ried J, Brown K, Piccolo FM, Cantone I, Landeira D, Bagci H, Hochegger H, Merkenschlager M, Fisher AG (2013) DNA synthesis is required for reprogramming mediated by stem cell fusion. Cell 152(4):873–883. https://doi.org/10.1016/j.cell.2013.01.012

14. Végran F, Apetoh L, Ghiringhelli F (2015) Th9 cells: a novel CD4 T-cell subset in the immune war against cancer. Cancer Res 75(3):475–479. https://doi.org/10.1158/0008-5472.CAN-14-2748

15. Odell ID, Cook D (2014) Optimizing direct immunofluorescence. Methods Mol Biol 1180:111–117. https://doi.org/10.1007/978-1-4939-1050-2_6

16. Donaldson JG (2001) Immunofluorescence staining. Curr Protoc Cell Biol. Chapter4:Unit 4 3. https://doi.org/10.1002/0471143030.cb0403s00

17. Sakaue-Sawano A, Kurokawa H, Morimura T, Hanyu A, Hama H, Osawa H, Kashiwagi S, Fukami K, Miyata T, Miyoshi H, Imamura T, Ogawa M, Masai H, Miyawaki A (2008) Visualizing spatiotemporal dynamics of multicellular cell-cycle progression. Cell 132(3):487–498. https://doi.org/10.1016/j.cell.2007.12.033

18. Whitfield ML, Sherlock G, Saldanha AJ, Murray JI, Ball CA, Alexander KE, Matese JC, Perou CM, Hurt MM, Brown PO, Botstein D (2002) Identification of genes periodically expressed in the human cell cycle and their expression in tumors. Mol Biol Cell 13(6):1977–2000. https://doi.org/10.1091/mbc.02-02-0030

19. Skarnes WC, Rosen B, West AP, Koutsourakis M, Bushell W, Iyer V, Mujica AO, Thomas M, Harrow J, Cox T, Jackson D, Severin J, Biggs P, Fu J, Nefedov M, de Jong PJ, Stewart AF, Bradley A (2011) A conditional knockout resource for the genome-wide study of mouse gene function. Nature 474(7351):337–342. https://doi.org/10.1038/nature10163

20. Yang J, Shen MH (2006) Polyethylene glycol-mediated cell fusion. Methods Mol Biol 325:59–66. https://doi.org/10.1385/1-59745-005-7:59

21. Gottesman A, Milazzo J, Lazebnik Y (2010) V-fusion: a convenient, nontoxic method for

cell fusion. BioTechniques 49(4):747–750. https://doi.org/10.2144/000113515

22. Hernandez JM, Podbilewicz B (2017) The hallmarks of cell-cell fusion. Development 144(24):4481–4495. https://doi.org/10.1242/dev.155523

23. Willkomm L, Bloch W (2015) State of the art in cell-cell fusion. Methods Mol Biol 1313:1–19. https://doi.org/10.1007/978-1-4939-2703-6_1

24. Kanda T, Sullivan KF, Wahl GM (1998) Histone-GFP fusion protein enables sensitive analysis of chromosome dynamics in living mammalian cells. Curr Biol 8(7):377–385. https://doi.org/10.1016/s0960-9822(98)70156-3

25. Kimura H, Cook PR (2001) Kinetics of core histones in living human cells: little exchange of H3 and H4 and some rapid exchange of H2B. J Cell Biol 153(7):1341–1353. https://doi.org/10.1083/jcb.153.7.1341

Chapter 10

Knockdown of Target Genes by siRNA In Vitro

Songhee Back and James J. Manfredi

Abstract

RNA interference (RNAi) is a cellular process involved in the silencing of genes, which makes RNAi important for observing and understanding the function of specific gene products. Short interfering RNA (siRNA) pathway is a RNAi pathway, where exogenous double stranded RNA is introduced to the cell and cleaved by an endoribonuclease, Dicer, to form siRNA, which interacts with a protein complex to scan mRNAs to bind to its complementary sequence. The binding of the siRNA to its complementary mRNA, the mRNA is cleaved and degraded by the cell, significantly reducing the levels of the target protein product. The discovery of this mechanism made it a powerful tool to use as a technique for therapeutics, agricultural biology, and cellular and molecular biology.

Key words RNAi, siRNA, Gene silencing, Gene knockdown, Cell cycle regulation, p53

1 Introduction

RNA interference (RNAi) is a biological process where double-stranded RNAs inhibit the expression of its target genes. It was first defined as posttranscriptional gene silencing (PTGS) when the overexpression of transgenes showed silencing of itself and the endogenous genes with sequence similarities. The mechanism behind PTGS was later shown to involve double-stranded RNA that significantly downregulated gene expression [1, 2]. These double stranded RNAs are cleaved by and endoribonuclease, Dicer, where the resulting short interfering RNA (siRNA), of about 25 nucleotide in length [3, 4], bind to the RISC complex and scans for its complementary mRNA sequence. The mRNA complementary to the siRNA is cleaved and degraded by the cell [5]. The elucidation of this pathway became a useful tool for cell biology, therapeutics, and agricultural biology [5, 6].

In cell biology, double-stranded RNA can be injected into mammalian cell cultures to knock down the expression of its target sequence mRNA and thereby significantly decrease the levels of protein product in the cell [4]. The knockdown of p53 can be used to help understand its role in cell cycle regulation. In response to stress, p53, a transcription factor, regulates the expression of genes involved in various cellular processes [7–9]. Some of these p53 responses include apoptosis, cell cycle arrest, and senescence, which are three of the most studied p53 regulated processes [7]. Since p53 is involved in preventing the progression of cell proliferation, elucidating the role of p53 is important in understanding cell cycle regulation, which is usually impaired in disease models such as tumorigenesis [8, 10] and neurodegenerative diseases [11]. In addition to p53, other players of the cell cycle can be knocked down to further shed light on the regulation of cell cycle and to develop therapeutics for cancer and neurodegenerative diseases.

2 Materials

1. On-TARGET plus siRNA or gene of interest (20 μM).
2. Allstars Negative Control siRNA (20 μM).
3. Oligofectamine™ Transfection Reagent.
4. Opti-MEM™|Reduced Serum Medium.
5. DMEM.
6. DMEM with 20% FBS: For 50 ml solution, 40 ml DMEM and 10 ml FBS.

3 Methods

All steps must be done under the tissue culture hood (*see* **Note 1**).

For one 100 mm dish (or 2 × 60 mm dishes) → 0.1 μM of siRNA per 100 mm dish (*see* **Notes 2–3**):

1. Make siRNA mix in Opti-MEM solution in falcon tubes. Per siRNA: 15 μl of siRNA (20 μM), 540 μl of Opti-MEM. Be sure to have one mix with negative control siRNA, so there must be at least two mixes made (*see* **Notes 4–5**).
2. Make Oligofectamine mix in Opti-MEM solution: 12 μl of Oligofectamine, 33 μl of Opti-MEM. Multiply the volumes by the number of mixes made in **step 1**.
3. Let the two mixtures sit in room temperature for 15 min.
4. Take the Oligofectamine mix (per 1 × 100 mm dish volume = 45 μl) and pipette mix into the siRNA mix.
5. Let the mix sit in room temperature for 15 min.

6. While the siRNA+Oligofectamine mix is sitting in room temperature for 15 min, wash the cells three times with DMEM without FBS (or 2× PBS and last time with DMEM without FBS).
7. Add DMEM (without FBS) into the siRNA–Oligofectamine mix so that 3 ml of the mixture is added to each 100 mm dish (for 1 × 100 mm dish, add 2.380 ml DMEM to the 620 μl siRNA+Oligofectamine mix).
8. Add 3 ml of the final mix (siRNA + Oligofectamine + Opti-MEM + DMEM) to each 100 mm dish (or 1.5 ml of mix to each 60 mm dish). This brings the final siRNA concentration per dish to 0.1 μM.
9. Incubate cells at 37 °C for 5 h.
10. Add 5 ml of DMEM with 20% FBS to each 100 mm dish (or 2.5 ml to each 60 mm dish).
11. Incubate at 37 °C for 24 h for siRNA knockdown. After 24 h the cells will be ready for collection or further experimental procedures (*see* **Note 6**).

4 Notes

1. Be sure to have all calculations ready before the start of the knockdown.
2. Split cells 24 h before they are treated with siRNA.
3. Cells should be at about 50–75% confluency for the transfection.
4. Pipette mix the solutions and do not vortex. Or gently flip tubes up and down to mix the solutions.
5. For multiple gene knockdowns, be sure to keep the total siRNA concentration at 0.1 μM per dish. Divide the total volume of siRNA needed by the number of genes that need to be knocked down for the volume of siRNA needed per gene.
6. To confirm knockdown, use RT-PCR and Western blot or immunocytochemistry. RT-PCR will check if the target mRNA is still present (Fig. 1), but this does not confirm if the protein levels has decreased. Western blot (Fig. 2) and immunocytochemistry will check if the target protein product is present.

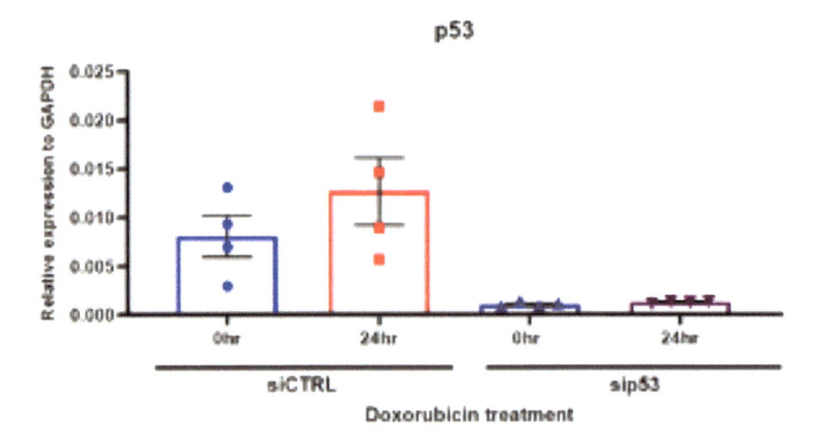

Fig. 1 Quantitative RT-PCR of p53 after knockdown and doxorubicin treatment in U2OS cells. U2OS cells were transfected with p53 siRNA or negative control siRNA and treated with 0.1 μM doxorubicin for 24 h. RNA levels determined by RT-PCR

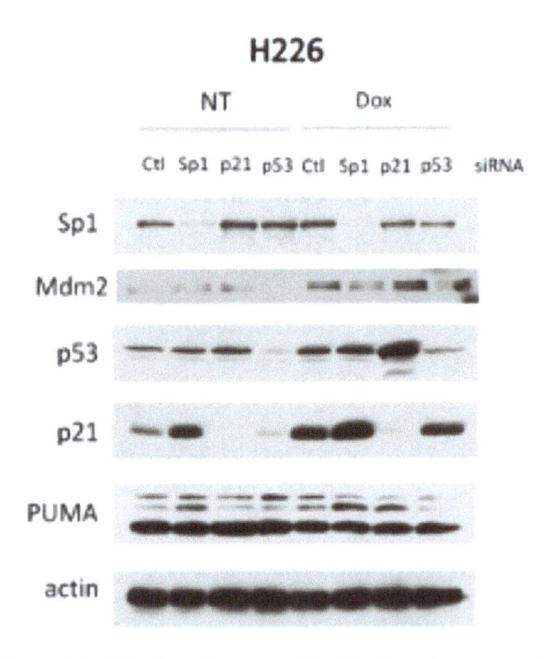

Fig. 2 Western blot of Sp1, p21, and p53 after knockdown and doxorubicin treatment. H226 cells were treated with each respective siRNA or negative control siRNA and treated with 0.1 μM doxorubicin for 24 h. Protein levels determined by Western blot

References

1. Fire A et al (1998) Potent and specific genetic interference by double-stranded RNA in *Caenorhabditis elegans*. Nature 391(6669):806–811
2. Fire A et al (1991) Production of antisense RNA leads to effective and specific inhibition of gene expression in *C. elegans* muscle. Development 113(2):503–514
3. Hamilton AJ, Baulcombe DC (1999) A species of small antisense RNA in posttranscriptional gene silencing in plants. Science 286(5441):950–952
4. Elbashir SM et al (2001) Duplexes of 21-nucleotide RNAs mediate RNA interference in cultured mammalian cells. Nature 411(6836):494–498
5. Dana H et al (2017) Molecular mechanisms and biological functions of siRNA. Int J Biomed Sci 13(2):48–57
6. Joga MR et al (2016) RNAi efficiency, systemic properties, and novel delivery methods for Pest insect control: what we know so far. Front Physiol 7:553
7. Vousden KH, Prives C (2009) Blinded by the light: the growing complexity of p53. Cell 137(3):413–431
8. Chen J (2016) The cell-cycle arrest and apoptotic functions of p53 in tumor initiation and progression. Cold Spring Harb Perspect Med 6(3):a026104
9. Sullivan KD et al (2018) Mechanisms of transcriptional regulation by p53. Cell Death Differ 25(1):133–143
10. Bieging KT, Mello SS, Attardi LD (2014) Unravelling mechanisms of p53-mediated tumour suppression. Nat Rev Cancer 14(5):359–370
11. Chang JR et al (2012) Role of p53 in neurodegenerative diseases. Neurodegener Dis 9(2):68–80

Chapter 11

Assaying Cell Cycle Status Using Flow Cytometry

Ramy Rahmé

Abstract

In this chapter, four methods are described that can be used to assess cell cycle status in flow cytometry. The first method is based on the simultaneous analysis of cellular DNA content using a fluorescent DNA dye (propidium iodide) and of a nuclear proliferation marker (Ki-67). The second is based on the differential staining of DNA and RNA using Hoechst 33342 and Pyronin Y: this method is particularly useful to distinguish quiescent cells in G0 phase from G1 cells. Finally, two methods are described based on DNA incorporation of the synthetic nucleosides BrdU and EdU.

Key words Cell cycle, Flow cytometry, Propidium iodide, Ki-67, Pyronin Y, Hoechst 33342, BrdU, EdU

1 Introduction

During cell cycle progression, proliferating cells undergo a sequential transition from G1 to S, S to G2, and finally G2 to M phases [1]. Besides, under certain circumstances, cells can enter G0 phase and stop dividing: this resting state is referred to as quiescence [2]. The simplest approach to analyze cell cycle status is to measure cellular DNA content using a fluorescent DNA dye. Based on this approach, cell cycle status can be classified in one of the three following distinct groups: G0/G1 (2n), S (2n to 4n), and G2/M (4n) [3].

Nevertheless, this method is insufficient to distinguish resting/quiescent cells (G0) from G1 phase cells. To overcome this limitation, fluorescent DNA dyes can be combined with proliferation-related markers such as Ki-67 and PCNA (Proliferating Cell Nuclear Antigen) in flow cytometry. For instance, Ki-67 antigen is rarely detected in G0 phase, highly expressed in the nuclear region of proliferating cells (maximum in G2 and early M phases) and rapidly degraded during subsequent phases of mitosis process (anaphase and telophase) [4]. Likewise, PCNA is a good marker for proliferating cells and is concentrated in S phase [5]. Alternatively, because highly proliferating cells usually contain higher levels of RNA compared to resting/quiescent cells, quantification of intracellular RNA by Hoechst 33342/Pyronin Y double staining is another approach mainly used to separate cells in G0 from cells in G1 [3, 6]. In the first part of this chapter, we describe basic flow cytometry techniques that are used to assess cell cycle status: the first protocol is based on DNA quantification and Ki-67; the second is based on quantification of intracellular RNA.

Cell cycle status can also be studied by measuring the amount of newly synthetized DNA content. Genomic DNA in proliferating cells can be labeled in vitro or in vivo in mice after exposing cells to BrdU (5′-bromo-2′-deoxyuridine), a thymidine analog, during the S phase of cell cycle. Incorporated BrdU is further stained with fluorochrome-conjugated anti-BrdU antibodies. Costaining with a fluorescent DNA dye (e.g., PI, 7-AAD) is used to separate the cells according to their cell cycle status (i.e., G1, S, G2/M) [7]. The main disadvantage of BrdU incorporation method is that both membrane permeabilization and harsh DNA denaturation are required for anti-BrdU antibody penetration.

As an alternative of BrdU, EdU (5-ethynyl-2′-deoxyuridine) was developed to overcome those limitations [8–10]. EdU is also a thymidine analog that carries a terminal alkyne group instead of a methyl in the [5] position of the pyrimidine ring. This alkyne group is exposed in the major groove of the DNA helix and can react with an organic azide in the presence of catalytic amounts of Cu(I) (e.g., Alexa 568-azide, TMR-azide—TMR-azide is cell-permeable, allowing cells to be stained while alive, without fixation and/or permeabilization): this "Click reaction" results in the formation of a stable triazole ring between EdU and the fluorescent azide. Therefore, EdU method is highly sensitive and faster than the classical BrdU method. In addition, EdU incorporation can be combined with multiple cell surface and/or intracellular staining protocols [10, 11]. The original version of the "Click reaction" cannot be used for multiplex detection of some fluorophores such as GFP and R-PE which are easily damaged by high concentration of copper and reactive oxygen species. Recently, chemical modification of "Click reaction" enables to preserve GFP and R-PE fluorescence and to obtain a bright EdU signal. This is extended to cover at least three different fluorophores (Alexa Fluor® 488, Alexa Fluor® 647, and Pacific Blue™). In the second part of the chapter, we describe BrdU- and EdU-based methods to study cell cycle.

2 Materials

Ki-67 is a nuclear antigen only present in proliferating cells (G1, S, G2, and M phases), but is absent from resting/quiescent cells (G0) [13, 14]. The Ki-67 antibody can detect the two Ki-67

2.1 Analysis of Cell Cycle Status by Costaining for DNA Content and Ki-67 [12]

isoforms of 320 and 359 kDa [15]. Staining for Ki-67 can be combined with a DNA dye, such as propidium iodide (PI), to assess cell cycle. PI is not membrane-permeable; therefore, live cells need to be permeabilized before staining with PI.

2.1.1 Solutions and Reagents

1. 1× PBS (Phosphate buffered saline): Gibco™ #10010072.
2. Bovine serum albumin (BSA) (Sigma-Aldrich, catalog #A3803).
3. FACS buffer: 1× PBS + 0.5% BSA (bovine serum albumin) (*see* **Note 1**).
4. 70% Cold ethanol (−20 °C).
5. FITC-conjugated Ki-67 antibody (eBioscience, catalog #11-5698-82: 0.5 mg/mL, 0.25μg/test).
6. PI (Invitrogen, catalog #P3566: 1 mg/mL).
7. RNase A (Invitrogen, catalog #12091021: 20 mg/mL).
8. PI staining solution: 1× PBS + 0.5% BSA + PI 5–10μg/mL (5–10μL of 1 mg/mL PI stock per 1 mL) + 100μg/mL RNase (5μL of 20 mg/mL RNase stock per 1 mL) + 2 mM MgCl$_2$ (2μL of 1 M MgCl$_2$ stock per 1 mL). Prepare freshly and keep in the dark at 4 °C before use.
9. Corning™ Falcon™ Test Tubes with Cell Strainer Snap Cap (Catalog #352235).

2.1.2 Special Equipment

Flow cytometer equipped with a 488 nm blue laser and appropriate filter sets detecting FITC and PI fluorescence.

2.2 Staining of Cell Surface Markers and for Ki-67/PI

This method can be used for multiplex staining of surface proteins (e.g., SLAM surface markers for hematopoietic stem cells) and Ki-67/DNA. After first staining of surface antigen, both fixation and permeabilization are subsequently required. We describe a classical procedure of surface marker staining followed by fixation with paraformaldehyde (PFA) and permeabilization with saponin. The saponin-based permeabilization is a reversible process; therefore, the remainder of the procedure should be performed in presence of saponin.

2.2.1 Solutions and Reagents

1. FACS buffer: 1× PBS + 0.5% BSA.
2. Fixation solution: 2% PFA.
3. Permeabilization solution: 10 mM HEPES buffer, pH 7.2 (10μL of 1 M HEPES buffer per 1 mL of final solution) + 1% (w/v) saponin (100μL of 10% w/v saponin stock solution per 1 mL of final solution). Prepare freshly and store at 4 °C before use.

4. Saponin wash buffer: 10 mM HEPES buffer, pH 7.2 + 0.5% (w/v) saponin. Prepare freshly and store at 4 °C before use (*see* **Note 2**).

5. Purified Rat Anti-Mouse CD16/CD32 (Mouse BD Fc Block™), Clone 2.4G2 (BD Biosciences, catalog #553141: 0.5 mg/mL).

6. Fluorophore-conjugated antibody against cell surface antigen (*see* **Note 3**).

7. FITC-conjugated Ki-67 antibody (eBioscience, catalog #11-5698-82: 0.5 mg/mL, 0.25μg/test).

8. PI (Invitrogen, catalog #P3566: 1 mg/mL).

9. PI staining solution/saponin: saponin wash buffer + PI 5–10μg/mL + 100μg/mL RNase + 2 mM $MgCl_2$. Prepare freshly and keep in the dark at 4 °C before use.

2.2.2 Special Equipment

Flow cytometer equipped with a 488 nm blue laser and appropriate filter sets detecting FITC and PI fluorescence. Depending on the fluorophore for surface antigens, additional laser and filter sets are needed.

2.3 Analysis of Cell Cycle by Hoechst 33342 and Pyronin Y Staining [12]

This method is mainly used to separate the quiescent cells (G0 phase) from cells in the G1 phase [16]. In general, quiescent cells in the G0 phase have lower levels of RNA compared with proliferating cells. Hoechst 33342 is a cell-membrane permeant dye that exclusively reacts with double-stranded DNA, while Pyronin Y, also a cell-membrane permeant dye, can react with DNA, RNA, and mitochondrial membranes. In the presence of Hoechst 33342, the interaction of Pyronin Y with DNA is disrupted. Thus, Pyronin Y can mainly stain RNA, when used at low concentrations [6], which allows for the quantification of RNA amount. Accordingly, when cells are stained first with Hoechst 33342 and then with Pyronin Y, it is possible to distinguish DNA from DNA using simultaneous fluorescence emitted by both dyes. In addition, this method offers another advantage: when these two dyes are used at low concentrations, stained cells can survive and can be sorted and cultured successfully [17].

2.3.1 Solutions and Reagents

1. FACS buffer: 1× PBS + 0.5% BSA.

2. Hoechst 33342, 20 mM Solution in water (Abcam #ab228551) (molecular weight 561.93 g/mol).

3. Hoechst-33342 staining solution: FACS buffer + 2μg/mL Hoechst-33342. Prepare freshly and keep in the dark at 4 °C before use (*see* **Note 4**).

4. Pyronin Y (Abcam #ab146350) (molecular weight 302.80 g/mol).

5. Pyronin Y stock solution: 1 mg/mL (*see* **Note 5**).

Assaying Cell Cycle Status Using Flow Cytometry 169

2.3.2 Special Equipment

Flow cytometer equipped with both 355 nm UV laser and 532 nm green or 561 nm yellow-green laser to activate Hoechst 33342 and Pyronin Y, respectively. Appropriate filter sets are needed (*see* **Note 6**).

2.4 Analysis of Cell Cycle Using BrdU Incorporation-Based Method

Here, we describe a "classical" in vitro BrdU/PI staining assay.

2.4.1 Solutions and Reagents

1. BrdU (Sigma-Aldrich, catalog #B5002: molecular weight 307.098 g/mol).
2. PI (Invitrogen, catalog #P3566: 1 mg/mL).
3. RNase A (Invitrogen, catalog #12091021: 20 mg/mL).
4. Tween 20 (Thermo Fisher Scientific, catalog #BP337-500).
5. Triton X-100 (Sigma-Aldrich, catalog #T9284).
6. Bovine serum albumin (BSA) (Sigma-Aldrich, catalog #A3803).
7. Monoclonal anti-BrdU antibody produced in mouse (Sigma-Aldrich, catalog #B2531).
8. FITC-conjugated goat anti-mouse IgG (Sigma-Aldrich, catalog #F0257).
9. Sodium tetraborate ($Na_2B_4O_7 \cdot 10H_2O$) (Sigma-Aldrich, catalog #B9876).
10. 0.05% trypsin–EDTA (Life Technologies, Gibco®, catalog #25300-062).
11. Phosphate buffered saline (PBS).
12. Ethanol.
13. HCl 37%, density of 1.2 g/mL at 25 °C.

2.5 Analysis of Cell Cycle Using EdU Incorporation-Based Method

Here, we describe an EdU incorporation-based method using EMD Millipore/Sigma-Aldrich reagents.

2.5.1 Solutions and Reagents

1. DMSO (Thermo Scientific™, catalog #85190).
2. 0.05% trypsin–EDTA (Life Technologies, Gibco®, catalog #25300-062).
3. Bovine serum albumin (BSA) (Sigma-Aldrich, catalog #A3803).
4. Triton X-100 (Sigma-Aldrich, catalog #T9284).
5. EdU Cell Proliferation Assay, EdU-647, 17-10528):
   (a) EdU vial (EMD Millipore catalog No. CS219082).

(b) Buffer additive vial (EMD Millipore catalog No. CS219074).

(c) Reaction buffer (EMD Millipore catalog No. CS219085).

(d) Catalyst solution (EMD Millipore catalog No. CS219038).

(e) Dye-Azide (10 mM) (EMD Millipore catalog No. CS219083).

2.5.2 Prepare Stock Solutions (See **Note 7**)

1. Prepare a 10 mM stock solution of EdU.
   (a) Add 2 mL of DMSO to EdU vial and mix to completely dissolve the EdU.
   (b) Store any remaining solution at −20 °C. When stored as directed, stock solution is stable for up to 1 year.

2. Prepare a 10× stock solution of the buffer additive.
   (a) Add 2 mL of deionized water to each of the vials and mix to have the compound dissolved completely.
   (b) Store any remaining solution at −20 °C. This stock solution is stable for up to 6 months if stored as directed. If solution starts to develop a brown color, it has degraded and should be discarded.

3 Methods

3.1 Analysis of Cell Cycle Status by Costaining for DNA Content and Ki-67

3.1.1 Harvest, Fix, and Permeabilize Cells Using 70% Cold Ethanol

1. Plate cells at proper density. Cells should not be confluent at the time of cell harvest (≈80% confluence).

2. Harvest live cells (1–2 × 10^6 cells), wash with 0.5–1 mL 1× PBS and pellet cells by centrifuging for 5 min at 1500 rpm (250 × g) at 4 °C.

3. Remove supernatant and resuspend cells in 0.5 mL 1× PBS.

4. Add 4.5 mL of prechilled 70% cold ethanol (−20 °C) to the cell suspension in a dropwise manner while vortexing.

5. Incubate the fixed cells 1–2 h at −20 °C. Cells may be stored in ethanol for several weeks at −20 °C prior to antibody staining.

3.1.2 Stain Cells with Ki-67 Antibody and Fluorescent DNA Dye

1. Pellet cells by centrifuging for 5 min at 2000 rpm (440 × g) at 4 °C and remove ethanol.

2. Rinse cells twice with 5 mL FACS buffer by centrifuging for 5 min at 1500 rpm (250 × g) at 4 °C.

3. Remove supernatant and resuspend cells in FACS buffer (1 × 10^6 cells/100μL).

4. Add FITC-conjugated Ki-67 antibody (0.25μg/10^6 cells) (*see* **Note 8**).

5. Incubate on ice for 30 min in the dark.

6. Wash twice with 1 mL 1× PBS by centrifuging for 5 min at 1500 rpm (250 × g) at 4 °C.

7. Remove supernatant then add 0.5 mL PI staining solution and gently resuspend the pellet.

8. Incubate for 20 min at room temperature in the dark. Do not wash after this step.

9. Add 50µL of PFA 4% only if you need to store the cells for 24–48 h before performing flow cytometry.

10. Filter your samples on Corning™ Falcon™ Test Tubes with Cell Strainer Snap Cap before performing flow cytometry in order to avoid cell clumping.

11. Analyze by flow cytometry.

3.2 Staining of Cell Surface Markers and for Ki-67/PI

3.2.1 Stain Cell Surface Antigens with Fluorophore-Conjugated Antibodies

1. Harvest live cells (1–2 × 10^6 cells), wash with 0.5–1 mL 1× PBS and pellet cells by centrifuging for 5 min at 1500 rpm (250 × g) at 4 °C.

2. Remove supernatant and resuspend cells in 100µL FACS buffer + 1:100 Fc Block. Incubate on room temperature for 5 min.

3. Add 100µL FACS buffer + 1:100 fluorophore-conjugated antibody detecting cell surface marker (for optimal antibody dilution, refer to the manufacturer's instructions or perform a titration). Incubate on ice for 30 min in the dark. After this step, the rest of the procedure should be done in the dark.

4. Wash twice with 200µL FACS buffer by centrifuging for 5 min at 1500 rpm (250 × g) at 4 °C.

3.2.2 Fix and Permeabilize Cells for Intracellular Staining

1. Resuspend cells in 200µL PFA 2%. Incubate for 20 min at room temperature in the dark.

2. Add 5 mL 1× PBS and centrifuge for 5 min at 1500 rpm (250 × g) at 4 °C to remove fixative.

3. Resuspend cells in 200µL of permeabilization solution (10 mM HEPES, pH 7.2, 1% saponin). Incubate for 20 min at room temperature in the dark. After this step, 0.5% saponin should be present in all buffers because the saponin-based permeabilization is reversible.

4. Wash cells with 5 mL saponin wash buffer (10 mM HEPES, pH 7.2, 0.5% saponin) by centrifuging for 5 min at 1500 rpm (250 × g) at 4 °C.

3.2.3 Stain Cells with Ki-67 Antibody and Fluorescent DNA Dye

1. Resuspend cells in 100µL saponin wash buffer + FITC-conjugated Ki-67 antibody (0.25µg/10^6 cells) (see **Note 8**).

2. Incubate on ice for 30 min in the dark.

3. Wash twice with 1–5 mL saponin wash buffer by centrifuging for 5 min at 1500 rpm (250 × g) at 4 °C.

4. Remove supernatant then add 0.5 mL PI staining solution (with 0.5% saponin) and gently resuspend the pellet.
5. Incubate for 20 min at room temperature in the dark. Do not wash after this step.
6. Filter your samples on Corning™ Falcon™ Test Tubes with Cell Strainer Snap Cap before performing flow cytometry in order to avoid cell clumping.
7. Analyze by flow cytometry.

3.3 Analysis of Cell Cycle by Hoechst 33342 and Pyronin Y Staining

1. Harvest live cells (1–2 × 10^6 cells) and wash with 1–5 mL 1× PBS by centrifuging for 5 min at 1500 rpm (250 × g) at 4 °C.
2. Resuspend live cells in their prewarmed (37 °C) optimal culture medium at a concentration of 1 × 10^6 cells/mL (*see* **Notes 9** and **10**).
3. Add 100μL of Hoechst 33342 staining solution. Hoechst concentration may require optimization.
4. Mix well and incubate in a water bath preadjusted to 37 °C for 45 min in the dark. Mix every 15 min (*see* **Note 11**).
5. Add Pyronin Y directly to cells with a final concentration of 4μg/mL (1–5μg/mL). Mix well and incubate in the water bath at 37 °C for another 45 min in the dark. Transfer cells onto ice and protect samples from direct exposure to light.

Option 1—For cell analysis only: Cells can be directly analyzed by flow cytometry (no wash is needed). Filter your samples on Corning™ Falcon™ Test Tubes with Cell Strainer Snap Cap before performing flow cytometry in order to avoid cell clumping.
Option 2—If cell sorting is needed

6. Wash cells with cold PBS by centrifuging for 5 min at 1500 rpm (250 × g) at 4 °C.
7. Resuspend in PBS + 2% FBS (fetal bovine serum) at a higher cell concentration (up to 20 million cells per mL). Keep cells at 4 °C in the dark during cell sorting.

Option 3: If staining for cell surface markers is needed

8. Centrifuge cells for 10 min at 1500 rpm at 4 °C.
9. Remove the supernatant and wash once with cold PBS containing Hoechst 33342 and Pyronin Y at the optimized concentration used above (*see* **Note 12**). Centrifuge for 10 min at 1500 rpm (250 × g) at 4 °C.
10. Discard the supernatant leaving 50–100μL of medium depending on total number of cells.
11. Add 1× PBS + 1:100 Fc Bloc. Incubate for 5 min at room temperature in the dark.

12. Add fluorophore-conjugated antibodies to the cells. Mix well and incubate on ice for 30 min in the dark (*see* **Notes 13** and **14**).
13. Wash with cold PBS containing Hoechst 33342 and Pyronin Y as described above.
14. Centrifuge the cells for 10 min at 1500 rpm (250 × *g*) at 4 °C.
15. Discard the supernatant and resuspend cells in FACS staining buffer (PBS + 0.5% BSA).
16. Proceed to analysis or cell sorting. Filter your samples on Corning™ Falcon™ Test Tubes with Cell Strainer Snap Cap before performing flow cytometry in order to avoid cell clumping.

3.4 Analysis of Cell Cycle Using BrdU Incorporation-Based Method

1. Culture cells under optimum growing conditions.
2. Add BrdU at a final concentration of 30μM: BrdU is light sensitive and should be added in the dark. Incubate in the dark for 30–60 min (*see* **Notes 15** and **16**).
3. Remove BrdU media, rinse once with PBS then trypsinize and harvest cells.
4. Centrifuge for 5 min at 1500 rpm (250 × *g*) at 4 °C.
5. Fix cells with 70% cold ethanol.
   (a) Resuspend pellet in 300μL PBS, agitate gently.
   (b) Add 700μL of ice-cold 100% ethanol in a dropwise manner.
   (c) Mix gently with a 1 mL glass transfer pipette.
   (d) Incubate at −20 °C for 1 h. Samples can be stored in ethanol for several weeks at −20 °C.
   (e) Pellet cells by centrifuging for 5 min at 1500 rpm (250 × *g*) at 4 °C, then completely aspirate supernatant.
6. Cell permeabilization and DNA denaturation: Add 0.5 mL of 2 N HCl–0.5% Triton X-100, and incubate for 30 min at room temperature in the dark (*see* **Note 17**).
7. Pellet cells by centrifuging for 5 min at 1500 rpm (250 × *g*) at 4 °C, then completely aspirate supernatant.
8. Resuspend cells in 0.5 mL of 0.1 M sodium tetraborate (molecular weight = 381.37 g/mol).
9. Incubate for 2 min then pellet cells by centrifuging for 5 min at 1500 rpm (250 × *g*) at 4 °C.
10. Wash cells once with 150μL PBS–BSA 1%.
11. Resuspend cells in 50μL PBS–0.5% Tween 20–BSA 1%. Add 1μg/10^6 cells of monoclonal anti-BrdU antibody (mouse antibody), and incubate for 1 h at room temperature in the dark.

12. Pellet cells by centrifuging for 5 min at 1500 rpm ($250 \times g$) at 4 °C, then wash once with 150 µL PBS–BSA 1%.

13. Pellet cells and resuspend in 50 µL PBS–0.5% Tween 20–BSA 1%. Add 1 µg/10^6 cells of FITC-conjugated goat anti-mouse IgG and incubate for 30 min at room temperature in the dark.

14. Pellet cells, resuspend pellet in 0.5 mL PBS containing 10 µg/mL RNase A and 20 µg/mL PI stock solution, and transfer to FACS tubes.

15. Leave samples at room temperature for 30 min in the dark. Analyze by flow cytometry.

3.5 Analysis of Cell Cycle Using EdU Incorporation-Based Method

3.5.1 EdU Incorporation Procedure

1. Culture cells under optimum growing conditions (≈80% confluence) in a 96-well plate.

2. Prepare a 10× working solution of EdU in prewarmed medium from the 10 mM EdU stock solution previously prepared. A suggested starting concentration range is 10–20 µM final. For example, to apply 10 µM final EdU concentration, make a 100 µM EdU working solution by diluting 10 mM stock 1:100 in complete media.

3. Apply 10 µL of 100 µM working solution to each well containing 100 µL media.

4. Incubate the cells in the dark for 30–60 min. After this step, the rest of the procedure should be done in the dark.

5. Aspirate culture media and rinse cells once with 100 µL 1× PBS per well.

6. Trypsinize the cells with 0.05% trypsin–EDTA pH 7.0 to detach adherent cells. Add 100 µL complete media to stop the reaction, then transfer the cell suspension to a new 96-well U-bottom plate.

7. Pellet the cells by centrifuging for 5 min at 1500 rpm ($250 \times g$) at 4 °C.

8. Discard the culture media by carefully pipetting without disturbing the cell pellets. Proceed immediately to fixation and permeabilization steps.

3.5.2 Cell Fixation and Permeabilization

1. Fixation: Add 100 µL of 4% PFA in PBS to each well of the 96-well plate. Incubate for 15 min at room temperature.

2. Remove the fixation solution by centrifuging the plate at 1500 rpm ($250 \times g$) for 5 min, then carefully remove solution without disturbing the cell pellets.

3. Saturation: Wash cells twice with 200 µL PBS + BSA 0.5% by centrifuging the plate at 1500 rpm for 5 min, then carefully remove solution without disturbing the cell pellets.

4. Permeabilization: Add 200μL of 0.5% Triton X-100 in PBS to each well. Incubate for 20 min at room temperature. Carefully remove the permeabilization solution after centrifuging the plate.

5. Wash cells twice with 200μL PBS + BSA 0.5%.

6. Proceed to fluorescent EdU detection steps (*see* **Note 18**).

3.5.3 EdU Detection (the "Click Reaction")

For slides or 96-well plates, 100μL of the reaction cocktail per well is recommended.

1. Prepare the reaction cocktail as described in the following table (Table 1). Mix the ingredients in the order listed. Use the cocktail within 15 min of preparation.

2. Add 100μL of reaction cocktail to each well. Rock the plate gently to distribute the reaction cocktail evenly. Incubate for 30 min at room temperature. Protect from light.

3. Remove the reaction cocktail, wash the cells three times with 200μL PBS + BSA 0.5%. Be careful not to disturb cell pellets when removing wash solution.
   Optional: Proceed with nuclear staining (DAPI or Hoechst 33342) or antibody labeling.

   If no additional staining is desired, proceed with imaging and analysis.

3.6 Flow Cytometry

Resuspend cells in 200–500μL PBS + BSA 0.5%: pipet up and down to suspend cells and have a single cell suspension. Run samples on FACS machine and collect flow data with appropriate excitation and emission wavelengths (Table 2).

3.7 Dye Dilution Proliferation Assay

Dye dilution assays using membrane-permeable fluorescent dyes can also be used to assess cell proliferation. Carboxyfluorescein diacetate succinimidyl ester (CFDA-SE) is one of the widely used fluorescent dyes. It enters the cytoplasm where it is cleaved by intracellular esterase enzymes to form an amine-reactive product: Carboxyfluorescein succinimidyl ester (CFSE). CFSE can covalently binds to intracellular amino acids and produces a detectable fluorescence [18, 19]. This reaction results in extremely long-term retention of the fluorescent dye within the original cell. When the cell divides, each daughter cell theoretically retains half of the CFSE dyes. Thus, this method is applied to estimate the number of the generation after rapid cell proliferation. Usually, CFSE can be traced through 6 to 8 generations by flow cytometry. Similar to CFSE, other fluorescent dyes have been developed to encompass a broad range of excitation/emission spectrum and avoid overlap with fluorescent proteins or fluorochrome-conjugated antibodies. Furthermore, some dyes emit in channels where cells have less natural autofluorescence and can be detected up to ten generations

Table 1
Reaction cocktail

Reaction cocktail (100μL per well)		
Component	Catalog number	Volume
Deionized water	–	75.8μL
Reaction buffer (10×)	CS219085	10μL
Catalyst solution	CS219038	4μL
Dye-Azide (10 mM)	CS219083	0.2μL
Buffer additive (10×)	CS219074	10μL

Table 2
Excitation and Emission Settings

Excitation and Emission Settings			
Catalgo number	Azide Dye	Excitation	Emission
17-10525	6-FAM	496 nm	516 nm
17-10526	5-TAMRA	546 nm	579 nm
17-10527	5/6-Sulforhodamine 101	584 nm	603 nm
17-10528	Eterneon-Red 645	646 nm	662 nm

during cell proliferation. For this analysis, you can refer to CellTrace™ CFSE Cell Proliferation Kit for flow cytometry, catalog #C34570, Invitrogen™.

4 Notes

1. Sodium azide NaN_3 0.1% can be added to prevent microbial contamination. FACS buffer can be stored at 4 °C for up to 6 months.

2. Optimization of PFA (1–4%) and saponin (0.1–1%) concentrations may be needed.

3. Choose a fluorophore whose detection wavelength has minimum overlap with FITC and PI ones.

4. Titration may be needed (concentration range of Hoechst 33342: 1–10μg/mL). Higher concentrations of Hoechst 33342 are toxic to cells while lower concentrations will lead to less resolved cell cycle phases.

5. Titration may be needed (concentration range of Pyronin Y (PY): 1–5μg/mL). Optimal PY concentration is critical: at low concentrations <1μg/mL, PY does not stain RNA

stoichiometrically; at high concentrations, PY is toxic to cells and PY/RNA complex precipitates, which in turn quenches PY fluorescence.

6. Hoechst 33342 dye is excited at 350 nm and its blue emission can be recorded between 440 and 480 nm. Pyronin Y optimal excitation is at 550 nm and its optimal emission is around 575 nm. Hence, the ideal cell analyzer for this method would be a flow cytometer fitted with UV (350 nm) and yellow-green (561 nm) lasers. However, a standard flow cytometer fitted with a UV laser is also convenient as Pyronin Y/RNA complex is efficiently excited by the standard 488 nm blue laser most flow cytometers have, and generates strong signals. Pyronin Y can be detected through the same emission filters (band pass filters around 575 or 580 nm) normally used for phycoerythrin (PE) on standard flow cytometers.

7. Aliquots are recommended to avoid repeated freeze–thaw cycles.

8. Titration of FITC-conjugated Ki-67 antibody may be required to achieve the best quality of positive cell discrimination from negative cells.

9. The cell density is crucial to ensure an optimal dye uptake. Higher numbers of cells can be stained as long as the final cell density is kept at 1×10^6 per mL. Include two extra tubes of one million cells each for single positive controls (one for Hoechst 33342 only and one for Pyronin Y only).

10. Fixed cells can also be stained with Hoechst 33342 and Pyronin Y (see Section 3.1.1 for cell fixation with ice cold ethanol 70%). Stain cells using 0.5 mL of Hoechst/PY staining solution. The sample is kept in the dark for 20 min at room temperature. No washing step is necessary.

11. Hoechst uptake by live cells is an active process and should be performed under optimal culture conditions depending on cell type.

12. Concentration of both Hoechst 33342 and PY should be maintained in all washing and staining solutions during immunophenotyping staining to prevent efflux of the dyes. When working with mouse bone marrow cells, verapamil hydrochloride (50μM in DMSO or Ethanol), a MDR1 protein inhibitor, could be added to the cells alongside Hoechst 33342 and PY to block dye efflux; or to check that stem-cell-enriched side population (SP), which has strong dye efflux activity, is not affecting the DNA and RNA profiles.

13. Prepare extra tubes for single positive controls ± fluorescence minus one (FMO) controls if needed.

14. With the introduction of Brilliant Violet and Brilliant Ultra Violet fluorochromes (BD Biosciences and BioLegend) in addition to standard fluorochromes such as FITC, PE-Cy7, APC, and its conjugates, there are a few good choices of nonoverlapping fluorochromes to combine with Hoechst 33342 and PY staining without major compensation issues. A suitable viability dye could also be added to exclude dead cells such as 7-AAD, PI, or Topro-3.

15. BrdU concentration may need optimization (ranges from 10 to 100μM), as well as time of pulse (ranges from 15 min to 2 h).

16. BrdU staining can also be performed in vivo.
    (a) Weigh mice to be injected with BrdU. BrdU quantity to be injected is 1 mg of BrdU per 6 g of body weight. The stock BrdU solution is 10 mg/mL, so 100μL per 6 g body weight is required.
    (b) Inject the prepared BrdU solution i.p. (intraperitoneal) 24–72 h before time of euthanasia: the more quiescent the population, the more time is required to gain measurable incorporation. As an adjunctive to i.p. injection (or as an alternative), especially for longer labeling studies, BrdU may also be administered in the drinking water. BrdU drinking water (1 mg/mL) should be protected from light and replaced daily; 1% glucose can also be added to improve palatability. Care should be taken if periods of longer than 1 week of administration are used as BrdU can become toxic to animals and may adversely affect results [20–22].

17. Standard HCl solution is 37% with a density of 1.2 g/mL at 25 °C, which means that HCl concentration is 12 M. To obtain a 2 N final concentration, the dilution factor is 6.

18. TMR-azide is cell-permeable organic azide, allowing cells to be stained while alive, without the need for fixation and/or permeabilization.

References

1. Malumbres M, Barbacid M (2009) Cell cycle, CDKs and cancer: a changing paradigm. Nat Rev Cancer 9:153–166
2. Zetterberg A, Larsson O, Wiman KG (1995) What is the restriction point? Curr Opin Cell Biol 7:835–842
3. Darzynkiewicz Z, Juan G, Srour EF (2004) Differential staining of DNA and RNA. Curr Protoc Cytom. Chapter 7, Unit 7.3
4. Gerdes J et al (1984) Cell cycle analysis of a cell proliferation-associated human nuclear antigen defined by the monoclonal antibody Ki-67. J Immunol 133:1710–1715
5. Kurki P, Vanderlaan M, Dolbeare F, Gray J, Tan EM (1986) Expression of proliferating cell nuclear antigen (PCNA)/cyclin during the cell cycle. Exp Cell Res 166:209–219
6. Shapiro HM (1981) Flow cytometric estimation of DNA and RNA content in intact cells stained with Hoechst 33342 and pyronin Y. Cytometry 2:143–150

7. Rothaeusler K, Baumgarth N (2007) Assessment of cell proliferation by 5-bromodeoxyuridine (BrdU) labeling for multicolor flow cytometry. Curr Protoc Cytom. Chapter 7, Unit7.31
8. Salic A, Mitchison TJ (2008) A chemical method for fast and sensitive detection of DNA synthesis in vivo. Proc Natl Acad Sci U S A 105:2415–2420
9. Cavanagh BL, Walker T, Norazit A, Meedeniya ACB (2011) Thymidine analogues for tracking DNA synthesis. Molecules 16:7980–7993
10. Cappella P, Gasparri F, Pulici M, Moll J (2015) Cell proliferation method: click chemistry based on BrdU coupling for multiplex antibody staining. Curr Protoc Cytom 72:7.34.1–7.34.17
11. Diermeier-Daucher S, Brockhoff G (2010) Dynamic proliferation assessment in flow cytometry. Curr Protoc Cell Biol. Chapter 8, Unit 8.6.1-23
12. Kim KH, Sederstrom JM (2015) Assaying cell cycle status using flow cytometry. Curr Protoc Mol Biol 111:28.6.1–28.6.11
13. Gerdes J, Schwab U, Lemke H, Stein H (1983) Production of a mouse monoclonal antibody reactive with a human nuclear antigen associated with cell proliferation. Int J Cancer 31:13–20
14. Schwarting R, Gerdes J, Niehus J, Jaeschke L, Stein H (1986) Determination of the growth fraction in cell suspensions by flow cytometry using the monoclonal antibody Ki-67. J Immunol Methods 90:65–70
15. Schlüter C et al (1993) The cell proliferation-associated antigen of antibody Ki-67: a very large, ubiquitous nuclear protein with numerous repeated elements, representing a new kind of cell cycle-maintaining proteins. J Cell Biol 123:513–522
16. Eddaoudi A, Canning SL, Kato I (2018) Flow cytometric detection of G0 in live cells by Hoechst 33342 and Pyronin Y staining. Methods Mol Biol 1686:49–57
17. Chitteti BR, Srour EF (2014) Cell cycle measurement of mouse hematopoietic stem/progenitor cells. Methods Mol Biol 1185:65–78
18. Lyons AB (2000) Analysing cell division in vivo and in vitro using flow cytometric measurement of CFSE dye dilution. J Immunol Methods 243:147–154
19. Lyons AB, Blake SJ, Doherty KV (2013) Flow cytometric analysis of cell division by dilution of CFSE and related dyes. Curr Protoc Cytom. Chapter 9, Unit9.11
20. Rocha B et al (1990) Accumulation of bromodeoxyuridine-labeled cells in central and peripheral lymphoid organs: minimal estimates of production and turnover rates of mature lymphocytes. Eur J Immunol 20:1697–1708
21. Reome JB et al (2000) The effects of prolonged administration of 5-bromodeoxyuridine on cells of the immune system. J Immunol 165:4226–4230
22. Matatall KA, Kadmon CS, King KY (2018) Detecting hematopoietic stem cell proliferation using BrdU incorporation. Methods Mol Biol 1686:91–103

Chapter 12

Using TUNEL Assay to Quantitate p53-Induced Apoptosis in Mouse Tissues

Lois Resnick-Silverman

Abstract

Critical to tumor surveillance in eukaryotic cells is the ability to perceive and respond to DNA damage. p53, fulfills its role as "guardian of the genome" by either arresting cells in the cell cycle in order to allow time for repair of DNA damage or regulating a process of programmed cell death known as apoptosis. This process will eliminate cells that have suffered severe damage from intrinsic or extrinsic factors such as X-ray irradiation or chemotherapeutic drug treatments that include doxorubicin, etoposide, cisplatin, and methotrexate. Assays designed to specifically detect cells undergoing programmed cell death are essential in defining the tissue specific responses to tumor therapy treatment, tissue damage, or degenerative processes. This chapter will delineate the TUNEL (terminal deoxynucleotidyl transferase nick-end labeling) assay that is used for the rapid detection of 3′ OH ends of DNA that are generated during apoptosis.

Key words p53, DNA damage, X-ray irradiation, Doxorubicin, Apoptosis, TUNEL assay, Immunofluorescence

1 Introduction

p53, a well-known tumor suppressor, can induce apoptotic cell death in response to a variety of different stress stimuli, activation of oncogenes, and DNA damage [1–3]. Upon activation, p53 can activate the transcription of well over 500 genes either directly or indirectly, that regulate many diverse cellular processes that include cell cycle arrest, apoptosis, cellular senescence, and differentiation. Major pathways include cell cycle arrest, p21 [4, 5], GADD45 [6], 14-3-3 sigma [7], DNA repair (p53R2) and apoptosis, PUMA [8, 9], NOXA, BAX, and Apaf-1. The overall effect of p53 activation is to inhibit cell growth by inducing DNA damage checkpoints and thereby prevent the passage of cellular mutations to subsequent generations and maintain a role as "guardian of the genome" [10, 11]. p53-dependent apoptosis relies on the contribution of direct transcriptional induction of the proapoptotic BH3 only proteins—PUMA and NOXA—and transcriptionally independent activities of p53 [12]. Cofactors that include JMY [13], ASPP [14], and other family members (p63 and p73) [15] may regulate the apoptotic function of p53. Additionally, posttranslational modification of p53 itself such as phosphorylation and acetylation have been shown to regulate the ability of p53 to activate the expression of apoptotic genes [16]. Induction of PUMA and NOXA have proved critical for killing tumor cells by anticancer drugs that activate p53.

Cell cycle arrest in the G1 phase is mediated by p21, an inhibitor of G1 cyclin-dependent kinases within complexes of cyclin A/CDK2, cyclin E/CDK2, and cyclin D/CDK2. The subsequent maintenance of the Rb-E2F complex prevents cells that carry damaged DNA from entering S phase and undergoing replication [17, 18]. This period of arrest will allow for repair and prevent the propagation of mutations to daughter generations. Cells lacking p21 fail to undergo a G1 arrest remain capable of undergoing apoptotic death [19].

G2/M arrest, a second checkpoint, G2/M can be achieved through the induction of p53 target genes GADD45 and 14-3-3 σ. The former can destabilize cyclin B/cdc2 complexes and the latter can modify cyclin B so that it remains sequestered in the cytoplasm. The cooperative action of these proteins prevents the entry into mitosis. Unlike the G1 checkpoint, G2/M is not solely p53 dependent.

Apoptosis is an evolutionarily conserved cell death pathway that is characterized by morphological and biologic changes in the cell. Three main pathways (intrinsic, extrinsic, and perforin/granzyme) can mediate the ultimate activation of caspase 3, the executioner, responsible for the degradation of DNA and cytoskeletal structures.

Critical to apoptotic response is the ability to regulate multiple BCL-2 prosurvival proteins. BCL-2 associated protein X (BAX) and BH3 only proteins, BCL-2 interacting death domain agonist (Bid) and p53 upregulated modulator of apoptosis (PUMA) interact with multiple BCL-2 prosurvival proteins in a cell type- and context-specific manner [20]. This leads to BAX and/or BCL-2 agonist/killer (Bak) oligomerization, mitochondria membrane permeability, caspase activation, and ultimately death. The induction of BH3 only proteins is essential for p53-dependent apoptosis to occur [21, 22].

PUMA induction in response to DNA damage or oncogenic stress is strictly p53 dependent, as has been demonstrated in vitro, in tissue cultured cancer cells, HCT 116 that lack p53 or H1299 cells that express the E6 protein that downregulates p53. These cells fail to induce PUMA and fail to apoptose following DNA damage [21, 23]. In addition, cells derived from mice that are deficient in PUMA expression are very resistant to apoptotic stimuli [24]. One can imagine that PUMA could be a good candidate as a therapeutic target. On the one hand, tumor growth could be inhibited by the induction of programmed cell death, and on the other hand, one might envision inhibiting PUMA to curb unwanted apoptosis associated with tissue injury or neurodegeneration.

Various assays have been developed for detecting DNA damage or various steps leading to programmed cell death. Antibody based methods against proteins that are associated with DNA damage or cell death pathways (e.g., H2AX or caspase 3) can be used but are not a direct measure of DNA fragmentation. These immunological methods require optimization of the antibodies for high specificity and low background. In combination with assays that measure DNA degradation they are useful validators of apoptosis. Assays used to determine the endonucleolytic degradation of DNA include the DNA ladder assay [25], Comet [26] and TUNEL assays [27, 28]. The DNA ladder assay is a qualitative method that used to determine the hallmark feature of apoptosis, DNA fragmentation. It uses agarose based electrophoresis to discern the characteristic DNA "ladder" produced by nucleochromosomal cleavage. However, it suffers from providing no information in regard to histological location at the single cell level. The Comet assay is used to determine single and double strand breaks in single cells that are embedded and lysed in a gel. It is best suited for in vitro tissue cultured cells as it might be difficult to extract the nuclei from in vivo tissues. This assay does not provide information about the size of the fragments because the DNA is bound to the nuclear matrix as it migrates through the gel forming the characteristic "comet tail." The extent of DNA damage is based on the brightness and length of the tail.

DNA fragmentation is a benchmark of apoptosis. The following method, terminal deoxynucleotidyl transferase (Tdt) deoxyuridine triphosphate (dUTP) nick end labeling, (TUNEL) is an assay used to detect and quantitate apoptotic cells. DNA damage can generate double and single stranded breaks that yield 3' hydroxyl (OH) termini in cells. Normal proliferating cells will have a minor number of DNA 3' OH ends and do not pose a problem. The 3' ends of the damaged DNA can detected by using the recombinant Tdt enzyme to catalytically incorporate fluorescein-12-dUTP at the 3' OH DNA ends. The enzymatically labeled DNA can then be visualized by fluorescence microscopy or analyzed by flow cytometry. This system is attractive because it is nonradioactive and can be completed within 4–5 h. The reagents are included in commercially available kits. If one uses tissue sections and fluorescent microscopy, apoptosis can be visualized at the single cell level. Generally, this assay produces a very high signal with minimal background noise. The inclusion of proper negative and positive controls makes the analysis very reliable (Fig. 1).

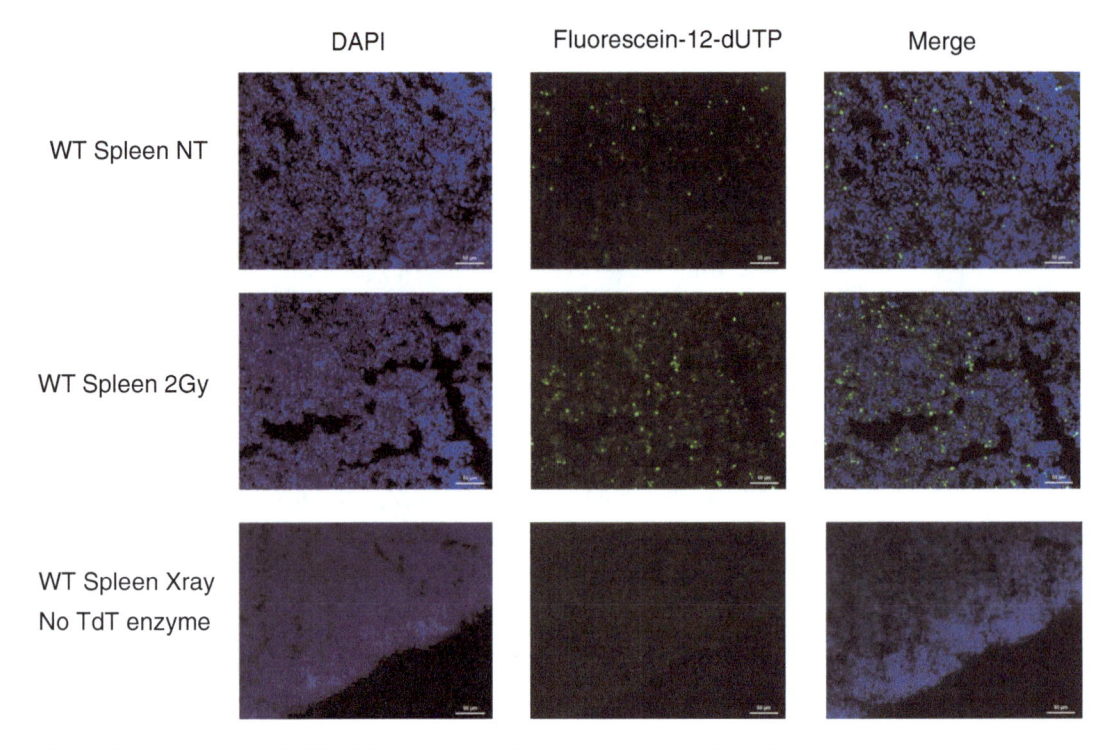

Fig. 1 X-ray treatment of p53 wild-type mice induces apoptosis in spleen tissue

2 Materials

2.1 Drug Treatment	1. Doxorubicin (Sigma).
2.2 Fixation of Mouse Organs	1. 4% Paraformaldehyde (Sigma P6148) in phosphate buffered saline (PBS) (*see* **Note 1**).
	2. 30% Sucrose in PBS.
	3. Optimal Cutting Temperature embedding medium (OCT) (Fisher 14-373-65).
	4. Disposable Base Molds (Fisher 22-363-553).
	5. Colorfrost Plus Precleaned Slides (Fisherbrand 22-230-891).
	6. ImmEdge Hydrophobic Barrier PAP Pen (Vector Laboratories H-4000).
2.3 Wash and Permeabilization	1. PBS.
	2. 0.2% Triton X-100 in PBS.
2.4 TUNEL Assay	1. DeadEnd Fluorometric TUNEL System (Promega G3250) The formulas for the following reagents in the kit (1a–1d) are the proprietary property of Promega.

(a) Equilibration Buffer.
(b) Nucleotide Mix.
(c) Terminal Deoxynucleotidyl Transferase (recombinant).
(d) SSC, 20×.
2. Plastic coverslips.
3. DNase Buffer containing 40 mM Tris 7.9, 10 mM NaCl, 6 mM $MgCl_2$, 10 mM $CaCl_2$.
4. Vectashield mounting medium containing DAPI (Vector Laboratories H1500).

2.5 Equipment

1. Cryostat.
2. X-ray irradiator.
3. Zeiss Axio Imager.Z1 with Fluorescence and Phase Contrast.

3 Methods

3.1 DNA Damage with Ionizing Radiation

1. Twenty-one-day-old C57/Bl6 mice are put in a Lucite pie shaped holder, placed on a revolving stand in the X-ray machine and irradiated with 2 Gy.
2. Mice are returned to holding cages. Age-matched mice that are not X-rayed are housed in similar cages.
3. At 3 h after X-ray treatment, both the irradiated and nonirradiated mice are sacrificed by regulated exposure to CO_2 for 4 min, a procedure that complies with the AVMA CO_2 Euthanasia Guidelines.
4. Dissection is carried out immediately following sacrifice.

3.2 DNA Damage with Doxorubicin

1. Twenty-one-day-old C57/Bl6 are weighed and injected intraperitoneally with 100 μl Doxorubicin in PBS (10 μg/g body weight).
2. Untreated control matched mice are injected with PBS.
3. Mice are returned to holding cages.
4. Mice are sacrificed after 48 h by regulated exposure to CO_2 for 4 min, a procedure that complies with the AVMA CO_2 Euthanasia Guidelines.
5. Dissection is carried out immediately following sacrifice.

3.3 Fixation of Mouse Organs

Day 1

1. Organs of interest are removed and placed in PBS that has been chilled on ice.

2. Organs are transferred into wells containing 4% paraformaldehyde in PBS and placed in the 4 °C cold room (*see* **Notes 1** and **2**).

Day 2

3. After 24 h, the 4%PFA/PBS is removed, replaced with 30% sucrose and returned to 4 °C (*see* **Note 3**).

Day 3

4. After another 24 h at 4 °C, the organs are ready for embedding in OCT (*see* **Note 4**).

5. Prepare a bed of dry ice.

6. Label molds and layer a small amount of OCT on the bottom of the mold.

7. The organ is placed in the center of the mold on the OCT. Mark the mold if orientation is important.

8. Cover the organ with OCT and fill to top of mold.

9. Place molds on dry ice in a level position until hardened.

10. Store molds at −80 °C until ready for sectioning.

3.4 Sectioning of Mouse Organs

1. Follow instructions on how to operate a cryostat or give molds to a core histology facility.

2. Cut 8–10 μM sections and lay at least 2 sections of organ on a charged microscope slide.

3. Store slides at −20 °C until day of assay.

3.5 TUNEL Assay

1. Remove sections from −20 °C freezer and allow slides to come up to room temperature on the benchtop until dry.

2. Apply a seal around the edges of the section using a hydrophobic marker and allow to dry.

3. Wash the slides twice with PBS (pH 7.4) at room temperature in order to rehydrate the sections. Add a sufficient volume to cover the section (*see* **Note 5**).

4. Permeabilize the sections with 0.2% Triton X-100 in PBS for 5 min at room temperature.

5. Rinse slides twice in PBS for 5 min at room temperature.

    (a) If you are including a positive control, treat the section with 100 μl of DNase buffer for 5 min at room temperature.

    (b) Remove the buffer and replace with 100 μl DNase buffer containing 5.5–10 units/ml of DNase for 10 min. at room temperature.

    (c) Remove the liquid and wash 3–4 times in deionized water in a Coplin jar that is only to be used for the positive control. Do not combine with other slides. Continue to **step 6** (*see* **Note 6**).

Please note that the following steps use reagents included in the Promega DeadEnd Fluorometric TUNEL System.

6. Cover all sections with 100 μl Equilibration Buffer for 5–10 min at room temperature.

7. As per kit instructions, all the rTdt labeling reagents are maintained on ice. Due to the extreme light sensitivity, these tubes should be covered in foil and slides are incubated in a dark chamber. A master mix can be assembled which provides that each slide will be incubated with 51 μl of Tdt-mediated labeling buffer that includes 45 μl Equilibration Buffer, 5 μl of Nucleotide Mix, and 1 μl of rTdt enzyme. A negative control slide in which the Tdt enzyme is not added to the mix should be included in this plan.

8. Tap off the Equilibration Buffer and blot around the sections before adding 50 μl of rTdt incubation buffer. The positive control slide will also be incubated with this mix. The negative control slide will be incubated with the mix that does not contain the rTdt enzyme. Cover each slide with Parafilm strips (*see* **Note 7**).

9. Place the slides on wet paper towels in a covered container and incubate in a 37 °C humidified chamber for 1 h.

10. Terminate the reaction by removing the coverslips and dipping slides into a Coplin jar that contains 2× SSC for 15 min.

11. Wash slides in PBS for 5 min. Repeat three times.

12. Remove the PBS and apply a drop of antifade fluorescence mounting media (Vectastain) over the section. Wait 1 min before laying down the coverslip.

13. Gently press the coverslip down and wipe the edges with a Kimwipes to remove excess liquid.

14. Store samples overnight in the dark at −20 °C before analyzing at the microscope.

3.6 Acquisition of Image

1. Images may be acquired using a conventional fluorescent microscope using DAPI and FITC filters.

2. Use a 20× or 40× magnification lens to determine the quality of the section and interpretation of the stain and counterstain. The 10× lens may also be used for quantitation of the TUNEL-positive puncta.

3. Once you determine the appropriate exposure, maintain those settings for the complete set of slides. This will give you an accurate comparison of TUNEL positive cells between each slide. Do not use automatic exposure.

4. Take images of the entire section in an organized way so that they can be stitched together at a later time for completeness.

5. Images can be analyzed manually or using commercially available image analysis software.

4 Notes

1. Paraformaldehyde (PFA) is a flammable solid. It reacts violently with strong oxidizers, and it is incompatible with inorganic acids and alkalis. Laboratory safety guidelines should be followed when preparing this reagent for the fixation solution. Please note that PFA breaks down at temperatures above 70°. A liter of 4% PFA can be prepared according to Cold Spring Harbor Protocols (cshprotocols.cship.org).

2. This fixation protocol is applicable to soft tissues. Bone will require a decalcification step following treatment with PFA. Otherwise it will not be amenable for molecular experiments, histology or immunocytochemical assays.

3. Placing the fixed tissue in 30% sucrose is important for cryoprotection of the tissue. The organs should sink in the sucrose solution between 15 h and 24 h at 4 °C indicating that infiltration of the sucrose is complete.

4. OCT embedding medium is preferred over resin or paraffin wax. You can rapidly turn out delicate sections with minimal processing that show little deformation of the tissue. The low antigen masking makes the sections particularly sensitive for immunohistochemistry.

5. Do not pipet directly onto the section as this might dislodge the tissue. Add liquids within the hydrophobic barrier at the edge of your tissue.

6. All positive control slides that have been exposed to DNase should be isolated from the experimental slides and washed separately to prevent cross contamination.

7. Using parafilm strips that are included in the kit or trimmed from rolls of laboratory parafilm will keep your sample from drying out for incubations greater than an hour. It will also greatly conserve the volume of reagents being used. Lay the strip over your sample at one edge and lower being sure not to trap any air bubbles.

Acknowledgments

The author would like to thank Nicolas Barthelery for his guidance in animal care and related procedures, as well as the use of the cryostat and sectioning instructions.

References

1. Kern SE (1991) Identification of p53 as a sequence-specificDNA-binding protein. Science 252(5013):1708–1711
2. Pietenpol JA, Tokino T, Thiagalingam S, El-Deiry WS, Kinzler KW, Vogelstein B (1994) Sequence-specific transcriptional activation is essential for growth suppression by p53. Proc Natl Acad Sci U S A 91:1998–2002
3. Burns TF, El-Deiry WS (1999) The p53 pathway and apoptosis. J Cell Physiol 181 (2):231–239
4. El-Deiry WS, Tokino T, Velculescu VE, Levy DB, Parsons R, Trent JM, Lin D, Mercer WE, Kinzler KW, Vogelstein B (1993) WAF1, a potential mediator of p53 tumor suppression. Cell 75(4):817–825
5. Harper JW, Adami GR, Wei N, Keyomarsi K, Elledge SJ (1993) The p21 Cdk-interacting protein Cip1 is a potent inhibitor of G1 cyclin-dependent kinases. Cell 75(4):805–816
6. Kastan MB, Zhan Q, El-Deiry W, Carrier F, Jacks T, Walsh WV, Plunkett BS, Vogelstein B, Fornace AJ (1992) A mammalian cell cycle checkpoint pathway utilizing p53 and GADD45 is defective in ataxia-telangiectasia. Cell 71:587–597
7. Chan TA, Hermeking H, Lengauer C, Kinzler KW, Vogelstein B (1999) 14-3-3Sigma is required to prevent mitotic catastrophe after DNA damage. Nature 401(6753):616–620
8. Nakano K, Vousden KH (2001) PUMA, a novel proapoptotic gene, is induced by p53. Mol Cell 7(3):683–694
9. Yu J, Zhang L, Hwang PM, Kinzler KW, Vogelstein B (2001) PUMA induces the rapid apoptosis of colorectal cancer cells. Mol Cell 7 (3):673–682
10. Sancar A, Lindsey-Boltz LA, Unsal-Kacmaz K, Linn S (2004) Molecular mechanisms of mammalian DNA repair and the DNA damage checkpoints. Annu Rev Biochem 73:39–85
11. Agarwal ML, Agarwal A, Taylor WR, Stark GR (1995) p53 controls both the G2/M and the G1 cell cycle checkpoints and mediates reversible growth arrest in human fibroblasts. Proc Natl Acad Sci U S A 92(18):8493–8497
12. Vousden KH, Lu X (2002) Live or let die: the cell's response to p53. Nat Rev Cancer 2 (8):594–604
13. Shikama N, Lee CW, France S, Delavaine L, Lyon J, Krstic-Demonacos M, La Thangue NB (1999) A novel cofactor for p300 that regulates the p53 response. Mol Cell 4(3):365–376
14. Samuels-Lev Y, O'Connor DJ, Bergamaschi D, Trigiante G, Hsieh JK, Zhong S, Campargue I, Naumovski L, Crook T, Lu X (2001) ASPP proteins specifically stimulate the apoptotic function of p53. Mol Cell 8(4):781–794
15. Flores ER, Tsai KY, Crowley D, Sengupta S, Yang A, McKeon F, Jacks T (2002) p63 and p73 are required for p53-dependent apoptosis in response to DNA damage. Nature 416 (6880):560–564
16. Liu Y, Tavana O, Gu W (2019) p53 modifications: exquisite decorations of the powerful guardian. J Mol Cell Biol 11(7):564–577. https://doi.org/10.1093/jmcb/mjz060
17. Waldman T, Kinzler KW, Vogelstein B (1995) p21 is necessary for the p53-mediated G1 arrest in human cancer cells. Cancer Res 55 (22):5187–5190
18. Bunz F, Dutriaux A, Lengauer C, Waldman T, Zhou S, Brown JP, Sedivy JM, Kinzler KW, Vogelstein B (1998) Requirement for p53 and p21 to sustain G2 arrest after DNA damage. Science 282(5393):1497–1501
19. Deng C, Zhang P, Harper JW, Elledge SJ, Leder P (1995) Mice lacking p21CIP1/WAF1 undergo normal development, but are defective in G1 checkpoint control. Cell 82 (4):675–684
20. Desagher S, Osen-Sand A, Nichols A, Eskes R, Montessuit S, Lauper S, Maundrell K, Antonsson B, Martinou JC (1999) Bid-induced conformational change of Bax is responsible for mitochondrial cytochrome c release during apoptosis. J Cell Biol 144 (5):891–901. https://doi.org/10.1083/jcb.144.5.891
21. Yu J, Wang Z, Kinzler KW, Vogelstein B, Zhang L (2003) PUMA mediates the apoptotic response to p53 in colorectal cancer cells. Proc Natl Acad Sci U S A 100(4):1931–1936

22. Yee KS, Vousden KH (2008) Contribution of membrane localization to the apoptotic activity of PUMA. Apoptosis 13(1):87–95. https://doi.org/10.1007/s10495-007-0140-2
23. Jeffers JR, Parganas E, Lee Y, Yang C, Wang J, Brennan J, MacLean KH, Han J, Chittenden T, Ihle JN, McKinnon PJ, Cleveland JL, Zambetti GP (2003) Puma is an essential mediator of p53-dependent and -independent apoptotic pathways. Cancer Cell 4(4):321–328
24. Wang P, Yu J, Zhang L (2007) The nuclear function of p53 is required for PUMA-mediated apoptosis induced by DNA damage. Proc Natl Acad Sci U S A 104(10):4054–4059. https://doi.org/10.1073/pnas.0700020104
25. Facchinetti A, Tessarollo L, Mazzocchi M, Kingston R, Collavo D, Biasi G (1991) An improved method for the detection of DNA fragmentation. J Immunol Methods 136(1):125–131. https://doi.org/10.1016/0022-1759(91)90258-h
26. Collins AR (2004) The comet assay for DNA damage and repair: principles, applications, and limitations. Mol Biotechnol 26(3):249–261. https://doi.org/10.1385/MB:26:3:249
27. Gorczyca W, Bruno S, Darzynkiewicz R, Gong J, Darzynkiewicz Z (1992) DNA strand breaks occurring during apoptosis - their early insitu detection by the terminal deoxynucleotidyl transferase and nick translation assays and prevention by serine protease inhibitors. Int J Oncol 1(6):639–648. https://doi.org/10.3892/ijo.1.6.639
28. Gavrieli Y, Sherman Y, Ben-Sasson SA (1992) Identification of programmed cell death in situ via specific labeling of nuclear DNA fragmentation. J Cell Biol 119(3):493–501. https://doi.org/10.1083/jcb.119.3.493

Chapter 13

Generation and Analysis of dsDNA Breaks for Checkpoint and Repair Studies in Fission Yeast

Rohana Ramalingam and Matthew J. O'Connell

Abstract

Damage to DNA elicits both checkpoint and repair responses. These are complex events that involve many genes whose products assemble at lesions and form signaling cascades to recruit additional factors and regulate the cell cycle. The fission yeast *Schizosaccharomyces pombe* has proven to be an excellent model to study these events, and has led gene and pathway discovery efforts. Recent progress has involved a more detailed analysis of the earliest events at lesions, particularly double-stranded DNA breaks (DSBs). Here we describe several methods for the analysis of events at DSBs, both on the DNA and the recruitment of proteins to these lesions, using *S. pombe* as a model. However, each of these methods is easily applicable to any experimental system with minor modifications to the protocols.

Key words Double-stranded DNA break (DSB), Immunofluorescence, Quantitative PCR (qPCR), *Schizosaccharomyces pombe*, End resection

1 Introduction

DNA damage comes in many forms with varying pathological consequences. The cellular responses following detection of the primary lesions both ensures their repair, and halts cell cycle progression via checkpoint signaling to allow time for the repair to proceed. While repair processes are lesion specific, and in some cases cell cycle regulated, the checkpoint responses are more limited in scope [1].

In higher eukaryotes, checkpoint responses to DNA damage can be divided into p53-dependent and p53-independent responses. p53 signals checkpoint responses through its transcriptional control of genes, predominately *CDKN1A*, which encodes the cyclin-dependent kinase inhibitor protein p21 [2]. The p53-independent DNA damage checkpoint is a signaling cascade culminating in the activation of the effector protein kinase Chk1, which in turn regulates the mitotic cyclin-dependent kinase Cdc2 via its inhibitory Y15 phosphorylation effects, Wee1, and Cdc25 [3, 4].

The most pathological form of DNA damage is the breakage of both strands of the DNA duplex, a double-stranded DNA break (DSB). In the G1 phase of the cell cycle, DSBs are repaired via direct ligation of DNA ends through an error-prone process known as nonhomologous end joining (NHEJ) [5]. In S- and G2-phase, the presence of a sister chromatid enables error-free homologous recombinational repair. In this case, the primary DSB is resected in a 5′→3′ direction to create single-stranded DNA with a 3′-OH group, and a 5′-PO_4 at the single-stranded/double-stranded junction. This structure is both necessary for the strand invasion into the sister chromatid for repair, and also acts as the structure on which complexes of checkpoint proteins form, including ATR/ATRIP, 9-1-1, and BRCT-containing mediator proteins, which is necessary to activate Chk1 [3, 6].

It has been known for some time that the checkpoint and repair responses to DSBs share the common intermediate structure generated by DSB resection, comparatively little is known about the formation of the ssDNA at lesions, how their formation is regulated in time and space, and how epigenomic events might affect this biology. To move understanding of these events forward, analysis of events at global and site-specific DSBs is critical. The fission yeast *Schizosaccharomyces pombe* has proven to be an excellent model for higher eukaryotic cell cycle regulation, checkpoint signaling, DNA repair and chromatin biology, with the added bonuses of excellent genetics and a genome that is 20× smaller than that in humans. Here, we outline several methods to study DSBs in *S. pombe*, though the general principles and many of the experimental approaches can easily be adapted to any eukaryotic cell. Application of these methods enable resolution of events by direct physical methods, and thus greatly augment previous studies based on cellular survival and cell cycle progression studies.

2 Materials

2.1 Electrophoretic Detection of Processed DSBs

1. Mid-sized horizontal agarose electrophoresis cell (We use Bio-Rad Sub-Cell GT system, Catalog No. 1704468, but any similar system will suffice).

2. Zetaprobe GT membrane (Bio-Rad, Catalog No. 1620196); an equivalent charged nylon membrane will suffice.

3. Standard low EEO agarose for DNA gels.

4. TAE running buffer (40 mM Tris, 20 mM acetic acid, 1 mM EDTA).

5. A kit and ^{32}P-radionucleotide for synthesis of DNA probes.
6. 0.4 M sodium hydroxide.
7. 20× SSC (3 M sodium chloride, 0.3 M sodium citrate, pH 7).
8. Whatman 3 MM paper (Aldrich, Catalog No. WHA3030931).
9. Autoradiographic film or Phosphorimager and screen.

2.2 Induction of DNA Double-Stranded Breaks and Microscopic Visualization of Sites of Homologous Recombination

1. Alexa Fluor™ 488 goat anti-rabbit IgG (H+L) antibody (Life Technologies, Catalog No. A11008).
2. Abcam Rabbit polyclonal to Rhp51 (Catalog No. ab63799).
3. DuraPore Membrane Filters (Catalog No. HVLP04700) (*see* **Note 1**).
4. Source to induce DSBs—the protocol uses a UV Crosslinker, but the same method can be applied to sources of ionizing radiation.
5. Electron Microscopy Sciences 16% Paraformaldehyde Solution (w/v), Methanol and RNase Free (Catalog No. 15710).
6. 25% glutaraldehyde (Sigma, Catalog No. G5882).
7. Lysing Enzymes from *Trichoderma harzianum* (Sigma, Catalog No. L1412).
8. PEM: 100 mM PIPES (piperazine-*N,N'*bis[2-ethanesulfonic acid]), 1 mM EGTA, 1 mM MgSO$_4$.
9. PEMS: PEM + 1.2 M sorbitol.
10. PEMB: PEM + 1% BSA.
11. 1% Triton X-100.
12. Zymolyase 20T from *Arthrobacter luteus* (Sunrise Scientific, Catalog No. N0766391).

2.3 Induction of DNA Double-Stranded Breaks and Processing of Lesions

1. ACROS Organics™ Anhydrotetracycline hydrochloride (Catalog No. AC233131000) 20 mM solution prepared in DMSO.
2. CSE Buffer (20 M citrate phosphate pH 5.6, 40 mM EDTA-Na pH 8.0, 1.2 M sorbitol).
3. 70% ethanol.
4. Isopropanol.
5. 1× PBS.
6. Phenol–chloroform–isoamyl alcohol (Fisher Scientific, Catalog No. AC327115000).
7. 5 M potassium acetate.
8. 100 mg/mL RNase.
9. STOP Buffer (150 mM NaCl, 50 mM NaF, 10 mM EDTA, and 1 mM sodium azide).
10. 1× TE (10 mM Tris pH 8.0, 1 mM EDTA).
11. 5× TE (50 mM Tris pH 8.0, 5 mM EDTA) with 1% SDS.

12. Zymolyase 20T from *Arthrobacter luteus* (Sunrise Scientific, Catalog No. N0766391).
13. 2× SYBR-Green PCR mix—we use Life Technologies PowerUp SYBR Green Master Mix (Catalog No. A25742), though any equivalent should suffice.
14. qPCR machine (We use a Life Technologies Step One Plus, but any equivalent machine will suffice).

2.4 Measuring Protein Recruitment to DNA Double-Strand Break Via Chromatin Immunoprecipitation (ChIP)

1. 0.5 mm diameter glass beads (Biospec, Catalog No. 11079105).
2. Bioruptor or similar bath sonicator (We use a Diagenode Bioruptor Plus System (Catalog No. B01020001).
3. ChIP Elution Buffer: (10 mM Tris pH 8.0, 1 mM EDTA, and 1% SDS).
4. ChIP High Salt Buffer: 50 mM HEPES, pH 7.5, 500 mM NaCl, 1% Triton X-100, 0.1% sodium deoxycholate, and 1 mM EDTA) (*see* **Note 2**).
5. ChIP Lysis Buffer: 50 mM HEPES, pH 7.5, 140 mM NaCl, 1% Triton X-100, 0.1% sodium deoxycholate, and 1 mM EDTA).
6. ChIP Wash Buffer: 10 mM Tris, pH 8.0, 250 mM LiCl, 0.5% Nonidet P-40, 0.5% sodium deoxycholate, and 1 mM EDTA).
7. Dynabeads™ Protein G (Catalog No. 10004D).
8. 36.5% formaldehyde (Sigma, Catalog No. F8775).
9. GE Healthcare illustra™ PCR DNA and Gel Band Purification Kit (Mfr. No. 28-9034-70).
10. 2.5 M Glycine.
11. 26G hypodermic needles.
12. 5× Loading Buffer: (250 mM Tris pH 6.8, 10% SDS, 30% Glycerol, 0.02% Bromophenol Blue, and 5% 2-mercaptoethanol).
13. qPCR machine (We use a Life Technologies Step One Plus, but any equivalent machine will suffice).
14. Mini Bead Beater (Biospec, Catalog No. 607).
15. 1× PBS.
16. 100 mM phenylmethylsulfonyl fluoride (PMSF) in ethanol.
17. Protease Inhibitors (Sigma, Catalog No. P8340; stock is 1000×, use at 5×).
18. 2× SYBR-Green PCR mix—we use Life Technologies PowerUp SYBR Green Master Mix (Catalog No. A25742), though any equivalent should suffice.
19. 6 Tube Magnetic Separation Rack (New England Biolabs, Inc., Catalog No. S1506S).

3 Methods

3.1 Electrophoretic Detection of Processed DSBs

There are several electrophoretic methods for analysis of DSBs. At a genome-wide level, chromosomal fragmentation and subsequent repair can be directly visualized by Pulse-Field Electrophoresis of whole and fragmented chromosomes. This method is straightforward, but requires a large amount of specialist (and very expensive) equipment such as a Bio-Rad CHEF Mapper. Thus, we do not discuss the method here, but details and examples can be found in the literature [7–9]. Alternatively, processing of DSBs can be studied by regular Southern blotting, where resection destroys sites for restriction enzymes and thus alters the pattern of hybridization [10]. While simple, this method is somewhat limited as it does not indicate the extent to resection into the restriction fragment itself. Below we describe a method to assess DSB end resection at specific sites that also shows the extent of single-stranded DNA production [11].

1. Grow cells, induce DSBs as outlined below in immunofluorescence protocol (Subheading 3.2), and extract genomic DNA as outlined below in Subheading 3.3.

2. Depending on the locus of interest, digest 10 μg of DNA with the appropriate restriction enzymes. We find this works best if the restriction fragment under analysis is 3–5 kb.

3. Load digested DNA on duplicate 0.7% Agarose/TAE gels. Gel A gets 9 μg of DNA, Gel B gets 1 μg of DNA, and run until the loading dye is approximately three-fourths down the gel.

4. Transfer the gels to Zetaprobe membrane using capillary transfer. Gel A is transferred with 20×SSC buffer, Gel B is transferred with 0.4 M Sodium Hydroxide. The DNA must be fixed to Membrane A using a UV cross-linker or baking in a vacuum oven. The DNA on membrane B is fixed by the transfer buffer.

5. Prepare probe DNA either by PCR or excision from a plasmid, label according to the kit instructions, and hybridize to each membrane.

6. Wash membranes and detect signal by autoradiography or phosphorimaging.

7. Details for transfer, hybridization, and washing of membrane are provided with the Zetaprobe and other commercial membranes. Any signal from membrane A emanates from single-stranded DNA produced in the cell by resection of the DSB, and will migrate both larger and smaller than the predicted size of the restriction fragment. The signal from membrane B emanates from single-stranded DNA produced by the 0.4 M sodium hydroxide transfer solution.

3.2 Detection of Sites of Homologous Recombination Through Rad51 Immunofluorescence

When cells are induced with DNA damage, DNA repair proteins accumulate at the sites of DNA double-strand breaks and are referred to as foci. Foci of some proteins can be observed through microscopy and represents ongoing sites of homologous recombination. Many proteins can be detected as fusions with florescent proteins by direct microscopy, though this assumes that the fusion proteins are fully functional. Further, some proteins such as Rad52 remain at sites of DNA damage after their repair [12]. In this protocol, Rad51 foci are visualized as this is a protein that is inactivated by addition of a florescent tag, and occupies damaged loci from the time of ssDNA production though to resolution of recombination. The method below is based on widely used immunofluorescence protocols for *S. pombe* that were initially developed for staining microtubules [13], but has been applied to many antigens. Where possible, a strain deleted for the gene of interest should be including as a negative control.

1. Start with logarithmically growing cells.
2. For each sample, gather two 150 mL conical flasks. Take 50 mL of sample and transfer to one flask. This will be the unirradiated sample.
3. Add 50 mL of your media of choice to the other flask. This will contain irradiated cells.
4. Using vacuum filtration, filter the remaining 50 mL of culture onto a membrane disc. For 50 mL of sample, filter 12–14 mL of culture onto one filter disc with a total of 4 filters per sample.
5. Place each filter cell side up in a petri dish using tweezers.
6. Place the petri dishes inside UV irradiator removing the lids.
7. Irradiate discs at your preferred dosage (*see* **Note 3**).
8. After irradiation, place all filters inside conical flask from **step 3**. Swirl flask. Swirling will dislodge cells from filter into the media.
9. Incubate samples for 30 min (at your preferred temperature).
10. Add 10 mL of 16% formaldehyde and 400 μL of 25% glutaraldehyde to each sample.
11. Incubate samples for 30 min (at your preferred temperature).
12. Transfer samples to 50 mL conical centrifuge tubes.
13. Centrifuge samples at $1500 \times g$ for 2 min. Decant supernatant.
14. Resuspend sample in 10 mL of PEM.
15. Centrifuge at $1500 \times g$ for 2 min. Decant supernatant.
16. Repeat **steps 14** and **15** twice for a total of three washes. Experiment may be paused at this stage. Incubate samples at 4 °C.
17. Prepare 10 mL of PEMS solution with 10 mg lysing enzymes and 3 mg zymolyase for each sample.

18. Transfer 5 mL of each sample to a 15 mL conical centrifuge tube. Resuspend before transferring if cells have settled to the bottom.
19. Centrifuge at 1500 × g for 2 min. Decant supernatant.
20. Resuspend each sample with 10 mL PEMS/lysing enzymes/zymolyase solution.
21. Incubate at 37 °C for ~25 min. Observe cells. Some cells will be phase dark on a phase-contrast microscope. Do not digest cells for too long.
22. Centrifuge samples at 1500 ×g for 2 min. Decant supernatant.
23. Resuspend sample in 10 mL PEM.
24. Repeat **steps 22** and **23** twice for a total of three washes.
25. Resuspend each sample with 10 mL 1% Triton X-100. Incubate at room temperature for 30 s.
26. Wash three times with 10 mL PEM (**steps 22–24**) (*see* **Note 4**).
27. Resuspend each sample in 400 μL PEMB. Transfer to a microcentrifuge screw cap tube.
28. Put tubes on rotator (or any apparatus that agitates/rotates tubes so that cells do not settle) at 4 °C to block for at least 4 h.
29. Add anti-Rad51 primary antibody at a dilution of 1:400 (*see* **Note 5**).
30. Incubate samples on rotator overnight at 4 °C.
31. Wash 3× with 1 mL PEM (**steps 22–24**).
32. Resuspend each sample in 400 μL PEMB.
33. Place tubes on rotator to block for 30 min.
34. Add secondary antibody Alexa Fluor 488 at a 1:400 dilution.
35. Incubate on rotator for 4 h at 4 °C.
36. Wash 3× with 1 mL PEM (**steps 22–24**).
37. Place 10 μL of sample onto a slide.
38. Observe under microscope.

3.3 Induction of a DNA Double-Stranded Break and Processing of Lesions

Sunder et al. developed the I-PpoI homing endonuclease system to induce a site-specific DNA double-strand break (DSB) in *Schizosaccharomyces pombe* in the ribosomal DNA (rDNA) or, after selection for loss of the site in the 28S gene, at an engineered I-PpoI site integrated into the *lys1* gene [14]. Compared to previous methods, this system can induce a DSB in a shorter time frame (30–60 min) and with a higher cutting efficiency of over 90%. This is far more efficient than using the HO-endonuclease from *Saccharomyces cerevisiae* expressed from the *nmt1* promoter, which is both inefficient and takes a protracted time to induce expression, meaning that

DSBs are being formed and repaired in the same population of cells [11]. Although this I-PpoI protocol shows how to induce a DSB in *S. pombe*, it can theoretically be used for other systems as the I-PpoI site is conserved in the 28S rRNA gene of all eukaryotes.

3.3.1 Induction of DNA Double-Strand Break (DSB)

1. Start with logarithmically growing cells. For each strain, 100 mL of culture is needed.
2. Add 50 mL of culture into two 150 mL conical flasks.
3. Add 5 µM anhydrotetracycline to one flask. This will be the treated sample.
4. After 2 h, place a 10 µL sample onto a slide. Observe under microscope. Ensure anhydrotetracycline treated cells are elongated before proceeding.

3.3.2 DNA Purification (This Method Is Also Used for the Electrophoretic Analysis Outlined in Subheading 3.1)

1. After 2 h, transfer sample to a 50 mL conical tube.
2. Centrifuge at $1500 \times g$ for 2 min. Decant supernatant.
3. Resuspend in 7.5 mL cold STOP buffer.
4. Place samples on ice for 10 min.
5. Centrifuge at $1500 \times g$ for 2 min. Decant supernatant.
6. To wash, resuspend in 7 mL 1× PBS.
7. Centrifuge at $1500 \times g$ for 2 min. Decant supernatant.
8. Resuspend in 5 mL CSE buffer with 5 mg/mL Zymolyase.
9. Incubate at 37 °C for 2.5–4 h.
10. Centrifuge at $1500 \times g$ for 2 min. Decant supernatant.
11. Resuspend in 1 mL of 5× TE with 1% SDS. Transfer to a 1.5 mL tube.
12. Incubate at 65 °C for 1–1.5 h.
13. Add 318 µL of 5 M potassium acetate. Mix by inversion.
14. Place on ice for 1–1.5 h.
15. Centrifuge tubes for $16,100 \times g$ for 15 min at 4 °C.
16. Collect 1.2 mL of supernatant per sample. Split this between two 1.5 mL tubes.
17. Add 600 µL cold isopropanol to each tube.
18. Incubate at 4 °C overnight.
19. Centrifuge at $16,100 \times g$ for 15 s. Decant supernatant.
20. Centrifuge at $16,100 \times g$ for 15 s. Remove supernatant with pipette.
21. Resuspend in 1 mL of cold 70% ethanol. Invert to mix.
22. Centrifuge at $16,100 \times g$ for 1.5 min. Decant supernatant.
23. Centrifuge at $16,100 \times g$ for 30 s. Remove supernatant with pipette.

24. Resuspend in 750 μL of 1× TE with 2 mM RNase. Add solution to one tube and resuspend the pellet. Then, transfer the volume to the other tube to resuspend that pellet.
25. Incubate at 37 °C for 1–1.5 h.
26. Add 750 μL of phenol–chloroform–isoamyl alcohol.
27. Vortex for 20 s.
28. Centrifuge at 16,100 × g for 5 min.
29. Transfer 650 μL of supernatant to another 1.5 mL tube.
30. Add 650 μL of phenol–chloroform–isoamyl alcohol.
31. Vortex for 20 s.
32. Centrifuge at 16,100 × g for 5 min.
33. Collect 500 μL of supernatant.
34. Add 250 μL of 7.5 M ammonium acetate and 750 μL of cold isopropanol. Mix by inversion.
35. Place on ice for 15 min.
36. Centrifuge at 16,100 × g for 10 min at 4 °C.
37. Decant supernatant.
38. Add 1 mL of cold 70% ethanol. Mix by inversion.
39. Centrifuge at 16,100 × g for 2 min at 4 °C. Decant supernatant.
40. Centrifuge at 16,100 × g for 30 s at 4 °C. Remove supernatant with pipette.
41. Dry pellets in the hood for approximately 20 min.
42. Resuspend in 50 μL of 1× TE.
43. Analyze DNA concentration and quality with NanoVue or equivalent UV spectrophotometer.
44. Dilute DNA to 50 ng/μL for future experiments.

3.3.3 Measuring Efficiency of Break Via qPCR

The principle here is that DSB formation by I-PpoI will destroy the ability to PCR across that site. Therefore, forward and reverse primers are needed that span the I-PpoI recognition sequence. These need to be designed for each locus under analysis, and examples are given for the 28S rRNA gene, and for *lys1*.

1. Dilute DNA to 1 ng/μL.
2. Load 12.5 μL of SYBR Green, 4 μL of 3.75 μM forward/reverse primer mix, and 8.5 μL of sample DNA into a 96-well plate.
Forward Primer:
For 28S: CGCAATGTGATTTCTGCCCAGTG.
For *lys1*: GAAGGTTTTAGGCAGTTCGAAC.

Reverse Primer:

For 28S: GTGGGAATCTCGTTAATCCATTC.

For *lys1*: ACTTTCCACACCCTAACTGACA.

The same principle can be applied to any locus (*see* **Note 6**).

All samples were run in duplicate, and the average of the duplicates are used for further analysis.

3. Run samples in a qPCR machine in the following conditions: 95 °C for 10 min.

   40 cycles of 94 °C for 15 s, 60 °C for 30 s, 72 °C for 30 s.

   95 °C for 15 s.

4. Calculate percent of DNA cut by $100 - \left(\frac{1}{2^{(DNA_{TET} - DNA_{no\ TET})}}\right)$.

3.3.4 Measuring Resection Length Via qPCR

DNA Digestion

The principle here is that resection past a site for a restriction enzyme destroys the ability for the restriction enzyme to cleave the site. Therefore, a PCR product can only be obtained after digestion with these restriction enzymes if the DSB is resected past that site. For examples of specific enzymes and PCR oligos for both the 28S rDNA and the *lys1* locus, please *see* [15]. The same principle using endogenous restriction enzyme sites can be applied to any locus.

1. For each digestion (100 ng of DNA), there is one control of undigested DNA. This sample contains only DNA and buffer.

2. In each tube, add the appropriate amount of enzyme (20 units), buffer (10X), and water into PCR tubes for a total of 40 μL.

3. Incubate samples (preferably in thermal cycler) at 37 °C for 4 h, 95 °C for 5 min, and held at 4 °C. Incubating samples at 95 °C inactivates the restriction enzymes.

4. After the digestion is complete, dilute the samples by adding 60 μL of nuclease-free water.

5. Load 12.5 μL of SYBR Green PCR Mix, 4 μL of 3.75 μM forward/reverse primer mix, and 8.5 μL of sample DNA into a 96-well plate.

   All samples are run in duplicate, and the average of the duplicates are used for further analysis.

6. Run samples in a qPCR machine in the following conditions: 95 °C for 10 min.

   40 cycles of 94 °C for 15 s, 60 °C for 30 s, 72 °C for 30 s.

   95 °C for 15 s.

3.3.5 Analysis

1. Firstly, a standard curve is generated to calculate amount of DNA from C_t values. DNA amounts were calculated for both the no TET and TET treated samples.

2. Using the amount of DNA, percent resection is calculated using the following equation.

$$\%\text{resection} = 200 \left(1 - \left(\frac{\Delta\text{DNA}_{\text{digested}}}{\Delta\text{DNA}_{\text{control}}}\right)\right) \div \left(2 - \left(\frac{\Delta\text{DNA}_{\text{digested}}}{\Delta\text{DNA}_{\text{control}}}\right)\right)$$

3.4 Measuring Protein Recruitment to DNA Double-Strand Break Via Chromatin Immunoprecipitation (ChIP)

1. Carry out all steps on ice unless otherwise specified.

2. For each sample, the appropriate control is needed. If using an epitope tagged protein, use an untagged control. If using a polyclonal antibody, use a null allele. Finally, if no viable null allele is available, use a nonspecific isotype control antibody.

3. Start with 48.6 mL of logarithmically growing cells. Transfer to a 50 mL conical tube.

4. Add 1.36 mL of 36.5% formaldehyde.

5. Incubate for 20 min at room temperature on roller.

6. Add 2.5 mL of 2.5 M glycine.

7. Incubate for 5 min at room temperature on roller.

8. Centrifuge at 1500 ×g for 2 min. Decant supernatant.

9. To wash, resuspend in 10 mL 1× PBS.

10. Centrifuge at 1500 × g for 2 min. Decant supernatant.

11. Resuspend in 1 mL 1× PBS. Transfer to 1.5 mL screw cap tube.

12. Centrifuge at 16,100 × g for 1 min. Remove supernatant with pipette.

13. Snap freeze pellet in liquid nitrogen. Experiment may be paused at this stage. Incubate samples at −80 °C.

14. Prepare ChIP Lysis Buffer with protease inhibitors (PI, 1/200 dilution) and PMSF (1/100 dilution).

15. Add glass beads to each sample.

16. Add 400 μL ChIP lysis buffer+PI/PMSF.

17. Place samples in bead beater for 5 min. Place on ice for 1 min.

18. Repeat **step 17** twice. *During this step, cool down centrifuge.*

19. Heat a 26G hypodermic needle in a Bunsen burner flame, and immediately pierce the bottom of the screw cap microfuge tube.

20. Quickly place screw cap tube into a 1.5 mL tube.
21. Centrifuge at 1500 × *g* for 1 min.
22. Remove the screw cap tube.
23. Centrifuge at 16,100 × *g* for 10 min at 4 °C. Decant supernatant. *During this step, prepare the protein G Dynabeads.*
24. Resuspend in 1 mL of ChIP lysis buffer+PI/PMSF.
25. Centrifuge at 16,100 × *g* for 10 min at 4 °C.
26. Resuspend in 400 μL ChIP lysis buffer+PI/PMSF.
27. Sonicate samples for 7 cycles of 30 s ON and 30 s OFF on high power (*see* **Note 7**).
28. Centrifuge at 5000 × *g* for 5 min at 4 °C.
29. Transfer supernatant to a screw cap tube.
30. Take 5 μL extract/supernatant and add to 100 μL ChIP Elution Buffer (this is INPUT sample).
31. Take 20 μL extract and add to 5 μL 5× Loading Buffer (this is for Western blot).
32. Add antibody of choice to the remainder of extract (*see* **Note 8**).
33. Incubate for 1 h at 4 °C on mixer.
34. Add 20 μL washed protein G coated Dynabeads and incubate overnight at 4 °C.

3.4.1 Preparing Dynabeads

1. Twenty microliters of beads is needed for each sample. Take required amount of beads and add 1 mL ChIP high salt buffer in a 1.5 mL tube.
2. Incubate on mixer at room temperature for 10 min.
3. Quickly vortex, and then place tube on a magnetic rack.
4. Remove buffer using pipette.
5. Resuspend beads in 1 mL of ChIP high salt buffer.
6. Incubate on mixer at room temperature for 10 min.
7. Resuspend beads with the same amount of ChIP lysis buffer as in **step 1**.
8. Place on ice until ready to add to extract.

3.4.2 ChIP Washes/Elution

1. To wash samples, quickly vortex tube and then place on magnetic rack. Remove buffer with pipette. Resuspend in the following buffers:
   1 mL ChIP lysis buffer followed by 5 min on mixer at 4 °C (2×).
   1 mL ChIP high salt buffer followed by 10 min on mixer at 4 °C (2×).

1 mL ChIP wash buffer followed by 5 min on mixer at 4 °C (2×).

1 mL TE followed by 1 min on mixer at 4 °C.

2. Add 130 μL elution buffer at room temperature.
3. Incubate at 65 °C for 2 h. Also, incubate INPUT samples from **step 30** (Subheading 3.4).
4. After 10 min, vortex samples and remove 20 μL for Western.
5. After 2h incubation, vortex, spin briefly, place on magnetic rack.
6. Transfer 100 μL of supernatant to a new tube.
7. Follow instructions given with GE Purification Kit to purify DNA.
8. Elute DNA in 170 μL water (2× 85 μL).
9. Load 12.5 μL of SYBR Green, 4 μL of 3.75 μM forward/reverse primer mix for each locus under analysis, and 8.5 μL of sample DNA into a 96-well plate.
    All samples were run in duplicate, and the average of the duplicates are used for further analysis.
10. Run samples in a qPCR machine in the following conditions: 95 °C for 10 min.

    40 cycles of 94 °C for 15 s, 60 °C for 30 s, 72 °C for 30 s.

    95 °C for 15 s.

3.4.3 Analysis (See Note 9)

1. Examine the melting curve to ensure each sample has returned a single PCR product.
2. For each site that protein recruitment is measured, calculate the C_t (Input-IP) for your negative control. This is the difference in C_t values between Input and IP.
3. Calculate the C_t (Input-IP) for your sample of interest.
4. Calculate ddC_t by subtracting negative control C_t (Input-IP) from sample C_t (Input-IP).
5. Finally, calculate protein recruitment by calculating 2 to the power of ddC_t (2^{ddC_t}).
6. Results can be reported as fold induction, or as the input is 1/80th of the IP, you can calculate % of input by multiplying the ΔC_t by 80.

4 Notes

1. We use these filters for optimal results, as cells do not stick to other filters we have previously tried.

2. The deoxycholate we use is Fisher Scientific Deoxycholic Acid, Sodium Salt (Catalog No. BP349). We have tried deoxycholate from other vendors, but have had inferior results.

3. We have found a dosage of 100 J/m^2 is optimal to induce damage without killing the cells.

4. Take extra care handling samples from this point forward as cells can be lost when decanting supernatant.

5. Antibody concentration should be optimized to each antibody.

6. A DSB can be induced at any locus in the DNA as long as there is a recognition site (natural or inserted) at that locus, which can be recognized by a homing endonuclease.

7. Sonication has been optimized to our conditions. To optimize, we sonicated for varying number of cycles and confirmed the DNA fragments were 300–600 bp long by running on an agarose gel.

8. It is important to determine the correct titer for each antibody to maximize the signal-to-noise ratio. We achieved this by performing a twofold dilution series until we could no longer see an increase in fold enrichment.

9. Microsoft Excel can be used for the analysis of multiple samples.

References

1. Barnum KJ, O'Connell MJ (2014) Cell cycle regulation by checkpoints. Methods Mol Biol 1170:29–40. https://doi.org/10.1007/978-1-4939-0888-2_2

2. Giono LE, Manfredi JJ (2006) The p53 tumor suppressor participates in multiple cell cycle checkpoints. J Cell Physiol 209(1):13–20

3. O'Connell MJ, Cimprich KA (2005) G2 damage checkpoints: what is the turn-on? J Cell Sci 118(Pt 1):1–6

4. O'Connell MJ, Walworth NC, Carr AM (2000) The G2-phase DNA-damage checkpoint. Trends Cell Biol 10(7):296–303

5. Chang HHY, Pannunzio NR, Adachi N, Lieber MR (2017) Non-homologous DNA end joining and alternative pathways to double-strand break repair. Nat Rev Mol Cell Biol 18(8):495–506. https://doi.org/10.1038/nrm.2017.48

6. MacDougall CA, Byun TS, Van C, Yee MC, Cimprich KA (2007) The structural determinants of checkpoint activation. Genes Dev 21(8):898–903. https://doi.org/10.1101/gad.1522607

7. Yadav RK, Jablonowski CM, Fernandez AG, Lowe BR, Henry RA, Finkelstein D, Barnum KJ, Pidoux AL, Kuo YM, Huang J, O'Connell MJ, Andrews AJ, Onar-Thomas A, Allshire RC, Partridge JF (2017) Histone H3G34R mutation causes replication stress, homologous recombination defects and genomic instability in S. pombe. Elife 6:e27406. https://doi.org/10.7554/eLife.27406

8. Outwin EA, Irmisch A, Murray JM, O'Connell MJ (2009) Smc5-Smc6-dependent removal of cohesin from mitotic chromosomes. Mol Cell Biol 29(16):4363–4375. https://doi.org/10.1128/MCB.00377-09

9. Verkade HM, Bugg SJ, Lindsay HD, Carr AM, O'Connell MJ (1999) Rad18 is required for DNA repair and checkpoint responses in fission yeast. Mol Biol Cell 10(9):2905–2918

10. Yan Z, Xue C, Kumar S, Crickard JB, Yu Y, Wang W, Pham N, Li Y, Niu H, Sung P, Greene EC, Ira G (2019) Rad52 restrains resection at DNA double-Strand break ends in yeast. Mol Cell 76(5):699–711 e696. https://doi.org/10.1016/j.molcel.2019.08.017

11. Kuntz K, O'Connell MJ (2013) Initiation of DNA damage responses through XPG-related nucleases. EMBO J 32(2):290–302. https://doi.org/10.1038/emboj.2012.322

12. Bass KL, Murray JM, O'Connell MJ (2012) Brc1-dependent recovery from replication stress. J Cell Sci 125(Pt 11):2753–2764. https://doi.org/10.1242/jcs.103119

13. Hagan IM, Hyams JS (1988) The use of cell division cycle mutants to investigate the control of microtubule distribution in the fission yeast Schizosaccharomyces pombe. J Cell Sci 89(Pt 3):343–357

14. Sunder S, Greeson-Lott NT, Runge KW, Sanders SL (2012) A new method to efficiently induce a site-specific double-strand break in the fission yeast Schizosaccharomyces pombe. Yeast 29(7):275–291. https://doi.org/10.1002/yea.2908

15. Barnum KJ, Nguyen YT, O'Connell MJ (2019) XPG-related nucleases are hierarchically recruited for double-stranded rDNA break resection. J Biol Chem 294(19):7632–7643. https://doi.org/10.1074/jbc.RA118.005415

Chapter 14

Calreticulin Exposure in Mitotic Catastrophe

Lucillia Bezu, Oliver Kepp, and Guido Kroemer

Abstract

Mitotic catastrophe is a modality of cell death (or occasionally senescence) that occurs after cells enter, and fail to resolve, abnormal mitosis, for instance after DNA damage or perturbations of the cell cycle. Mitotic catastrophe can avoid the generation of neoplastic cells from premalignant precursors, yet may also occur in cancer cells as a result of radiotherapy or chemotherapy. Of note, vinca alkaloids and taxanes, which are both known for affecting the stability of microtubules, can trigger mitotic catastrophe. Such agents can also cause cancer cells to undergo immunogenic cell death (ICD), which allows therapeutic responses to last beyond treatment discontinuation due to the induction of an antitumor immune response. ICD is commonly characterized by the exposure of the endoplasmic reticulum protein calreticulin on the cell surface. Here we describe an immunofluorescence-based cytofluorometric technique to detect calreticulin exposure on tumor cells exposed to drugs that induce mitotic catastrophe.

Key words Calreticulin, Immunogenic cell death, Vinca alkaloids, Taxanes, Cytometry

1 Introduction

The term "mitotic catastrophe" describes the entry of cells into aberrant mitosis, leading to chromosome missegregation and failed cell division. Mitotic catastrophe can be triggered by physical stress that directly affects DNA integrity (e.g., ionizing radiation, hyperthermia) and chemical stress or pharmacological stress inflicted by microtubular poisons that disrupt or block mitotic spindles, thus inhibiting mitosis progression at the stage of the metaphase. This process leads to cell death by apoptosis or necrosis or to an irreversible cell cycle arrest termed senescence. If mitotic catastrophe does not occur, for instance due to the absence of caspase-2, cells that survive aberrant mitosis tend to accumulate multiple numeric and structural abnormalities, with the consequent risk of oncogenesis [1–8]. Thus, mitotic catastrophe can be viewed both as a tumor suppressor mechanism (protecting against oncogenesis occurring spontaneously or in response to carcinogens or genetic factors) and as the therapeutic endpoint of radiotherapy and chemotherapy [9, 10].

Prominent inducers of mitotic catastrophe are the microtubular poisons used as chemotherapeutic agents including vinca alkaloids (e.g., vincristine, vinorelbine) and taxanes (e.g., paclitaxel, docetaxel). These agents have previously been described as being capable of inducing a specific cell death modality that is called immunogenic cell death (ICD) [11]. In response to immunogenic chemotherapeutics, dying tumor cells emit signals consisting in the cell surface exposure or release of damaged-associated molecular patterns (DAMPs). Such DAMPs include the most abundant chaperone protein of the endoplasmic reticulum, calreticulin (CALR), which is exposed at the surface of dying cancer cells, as well as a number of factors that are secreted, namely, adenosine triphosphate (ATP), the cytoplasmic protein annexin A1 (ANXA1), the nuclear protein high mobility group box 1 (HMGB1), and type I interferons [12–16]. Extracellular ATP causes the attraction of myeloid cells including dendritic cell (DC) precursors into the tumor bed [17, 18]. ANXA1 facilitates the juxtaposition of immature DC to stressed and dying cancer cells. Within such agonizing tumor cells, phosphorylation of the eukaryotic translation initiation factor 2α (EIF2A, better known as eIF2α) by stress-elicited kinases ultimately causes CALR to translocate from the lumen of the endoplasmic reticulum to the outer surface of the plasma membrane, where it acts as an "eat-me signal" for DC [12, 19–26]. HMGB1 that binds to Toll-like receptor-4 on DCs increases their antigen presentation efficacy [27]. As a result, after phagocytosis of portions of dying cancer cells, mature DCs expose and cross-present tumor antigens to CD8+ lymphocytes (which are attracted into the tumor bed by type 1 interferon responses), hence igniting an immune response that ultimately leads to the destruction of residual and resistant tumor cells [16, 28–30].

Microtubular inhibitors efficiently trigger the translocation of CALR to the plasma membrane, which is one of the limiting factors of ICD [31]. Accordingly, these agents have been described to induce anticancer immune responses in mouse models [31–34]. It is important to note that hyperploidization triggered by microtubular poisons or inhibitors of the actin cytoskeleton is sufficient to stimulate eIF2α phosphorylation leading to CALR exposure [31]. This suggests that mitotic catastrophe, the normal fate of hyperploid cells, is accompanied by CALR exposure [35, 36].

Here we report an immunofluorescence technique for detecting CALR exposure on cells that are placed in conditions that may induce mitotic catastrophe. We provide the technical details to perform cytofluorometric quantitation of CALR exposure on the plasma membrane of tumor cells shortly after treatment with 4 different mitotic poisons (Figs. 1 and 2).

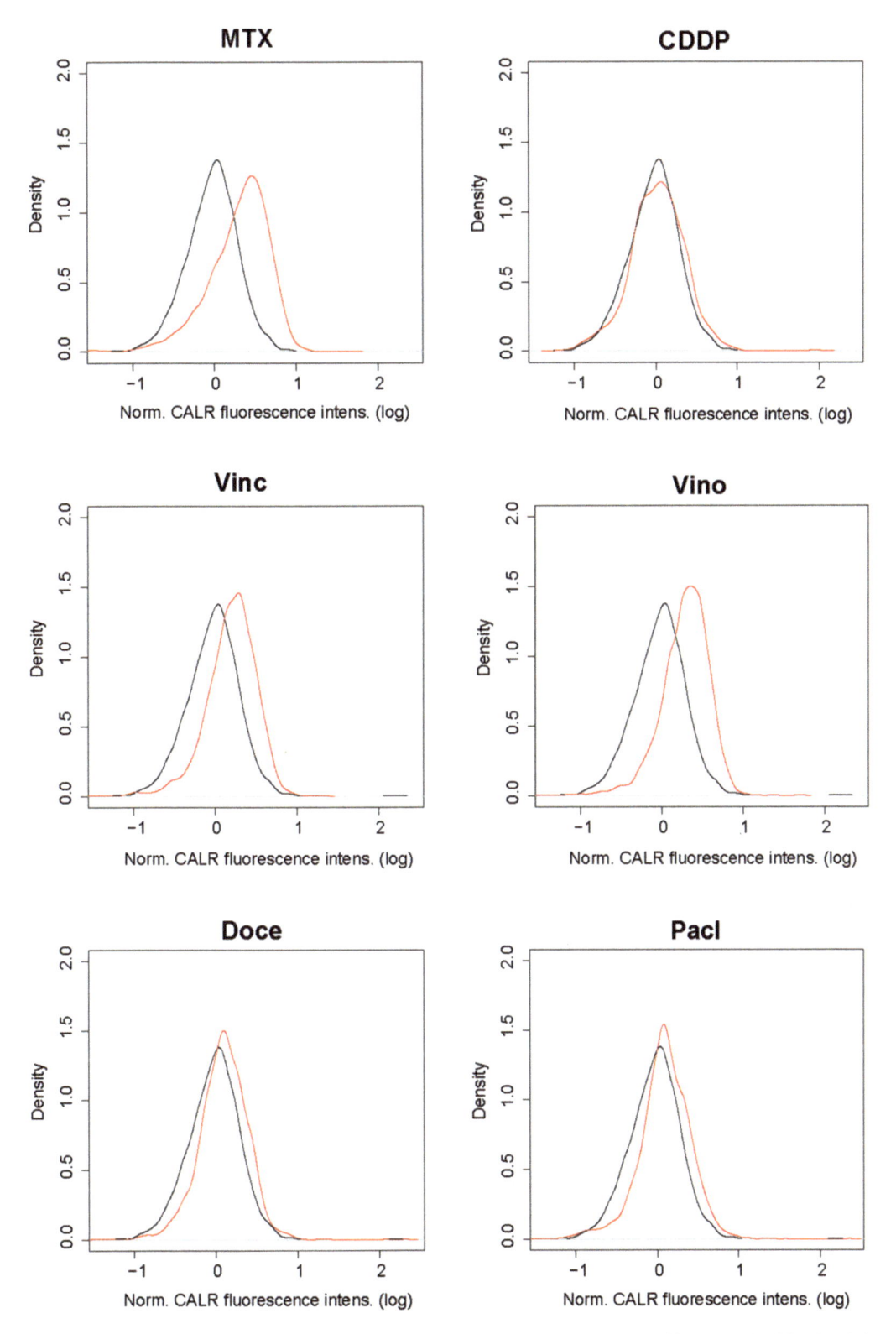

Fig. 1 U2OS cells were treated for 6 h with mitoxantrone (MTX, 3 μM), cisplatin (CDDP, 150 μM), vincristine (Vinc, 3 μM), vinorelbine (Vino, 3 μM), docetaxel (Doce, 3 μM), or paclitaxel (Pacl, 3 μM). Calreticulin (CALR) exposure was determined by immunofluorescence staining and flow cytometry and is expressed as normalized fluorescence intensity while excluding dead (DAPI-positive) cells. The black curve indicates the fluorescence of untreated controls, while the red curve corresponds to the treatment with the indicated drugs

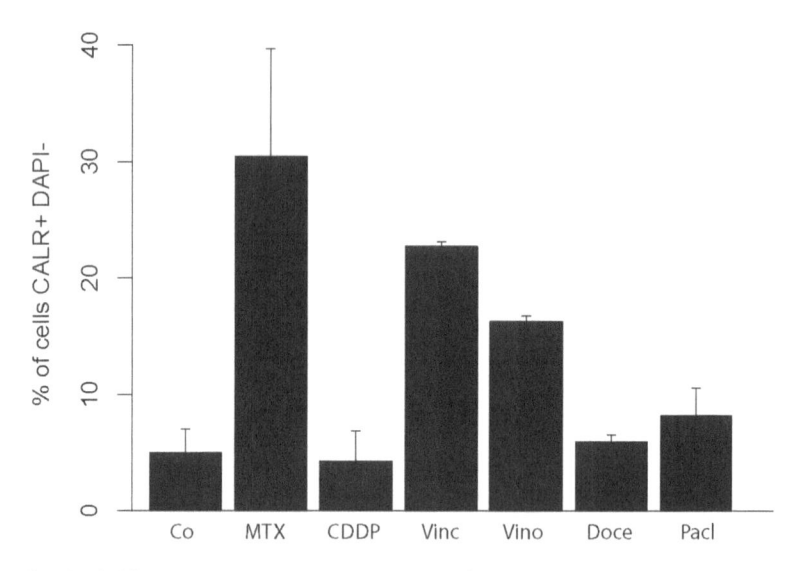

Fig. 2 U2OS cells were treated as in Fig. 1 for 6 h with mitoxantrone (MTX, 3 μM), cisplatin (CDDP, 150 μM), vincristine (Vinc, 3 μM), vinorelbine (Vino, 3 μM), Docetaxel (Doce, 3 μM), and Paclitaxel (Pacl, 3 μM). The results from Fig. 1 were expressed as the percentage of calreticulin positive, DAPI negative (CALR$^+$ DAPI$^-$) cells. Data are represented as the mean±SEM of triplicates from one representative out of three independent experiments

2 Materials

2.1 Disposable

1. 75 cm^2 flasks for cell culture.
2. Aspirating pipettes and serological pipettes (5 ml, 10 ml, and 25 ml).
3. Cell counting slides.
4. 1.5 ml eppendorf tubes.
5. Pipetting reservoirs.
6. Tips (1–20 μl, 20–200 μl, and 200–1000 μl).
7. Multichannel pipettes (20–200 μl).
8. 96-well plate (Greiner Bio-One; Kremsmünster, Austria).
9. 96-well plate V-shape (Greiner Bio-One; Kremsmünster, Austria).
10. Aluminum foil.

2.2 Equipment

1. Humidified incubator at 37 °C with 5% CO_2.
2. Centrifuge for 96-well plate under 4 °C.
3. Laminar flow hood.
4. Light microscope.

5. CyAn™ Advanced Digital Processing (ADP) cytofluorometer coupled to a HyperCyte loader.
6. R software for statistical analysis (*see* **Note 1**).

2.3 Reagents

1. Dulbecco's Modified Eagle's Medium (DMEM) supplemented with 10% fetal bovine serum, 1% nonessential amino acids, 1% HEPES buffer, and 1% penicillin/streptomycin sulfate.
2. Human osteosarcoma U2OS cells (*see* **Note 2**).
3. 0.05–0.25% trypsin solution with EDTA.
4. 1× phosphate-buffered saline (PBS) at pH 7.4.
5. Paraformaldehyde (PFA) at 3.7%.
6. Hoechst 33342 at 4 μM.
7. Bovine serum albumin (BSA) at 1% (dilution in PBS).
8. Fetal bovine serum (FBS).
9. Rabbit monoclonal antibody against calreticulin (Abcam #ab2907).
10. Alexa Fluor 488® goat anti-rabbit antibody.
11. 4′,6-diamidino-2 phenylindole dihydrochloride (DAPI) 5 mg/ml.
12. Drugs of interest: mitoxantrone (MTX), cisplatin (CDDP for cis-diamminedichloridoplatinum(II)), vincristine (Vinc), vinorelbine (Vino), paclitaxel (Pacl), and docetaxel (Doce).
13. Dimethyl sulfoxide (DMSO).

3 Methods

3.1 Cell Culture

1. Adherent cells have to be cultured for 1 week in complete DMEM.
2. Use cells at a maximum of 80% confluency.

3.2 Drug Preparation

1. Dilute the following drugs in DMSO and store them in Eppendorf tubes: MTX (positive control), vinca alkaloids (Vinc, Vino), and taxanes (Pacl, Doce). Dilute CDDP (negative control) in PBS.
2. Keep MTX and CDDP at 4 °C and protected from light exposure. Store vinca alkaloids and taxanes at −20 °C.

3.3 Cell Seeding

1. Seed 4000 U2OS cells in 100 μl DMEM per well in 96-well plates (*see* **Note 3**).
2. The following day, at 80% of confluence, treat the cells for 6 h with the drugs of interest (vinca alkaloids, taxanes), positive and negative control (MTX and CDDP respectively). Drugs have to be used at 3 μM, except CDDP, which is used at 150 μM (*see* **Note 4**).

3.4 Immunostaining

1. After the treatment, gently aspire the culture supernatant.
2. Add 100 μl of trypsin in each well. Wait for 4 min at 37 °C (*see* **Note 5**).
3. Add 50 μl of FBS in each well.
4. Transfer the mix of cells, trypsin and FBS in a V-shaped 96-well plate.
5. Centrifuge the cells for 5 min at 1200 rpm (400 × g) at 4 °C.
6. Carefully remove the supernatant. Add 20 μl of primary rabbit monoclonal antibody against calreticulin (1/100 diluted in BSA 1%) in each well. Vortex for 30 s in order to mix the cells with the antibody.
7. Incubate 30 min at 4 °C protected from light with an aluminum foil and keep the plate on ice.
8. Add 150 μl of PBS in each well and centrifuge 5 min at 1200 rpm (400 × g) at 4 °C.
9. Remove supernatant. Add 20 μl of secondary Alexa Fluor® 488 goat anti-rabbit IgGs (1/500 diluted in BSA 1%) in each well. Vortex for 30 s in order to mix the cells with the antibody.
10. Incubate 30 min at 4 °C protected from light with an aluminum foil and keep the plate on ice.
11. Add 150 μl of PBS in each well and centrifuge 5 min at 1200 rpm (400 × g) at 4 °C.
12. Remove supernatant. Add 75 μl of DAPI (1/2000 diluted in PBS). Vortex for 30 s in order to resuspend the cells (*see* **Note 6**).
13. Analyze the cells immediately with the cytometer.

3.5 Cytometer Preparation

1. Analyze samples using a CyAn™ ADP cytofluorometer coupled to a HyperCyt loader (*see* **Note 7**).
2. Empty the waste container and refill the containers for distilled water and ethanol (EtOH) 70%.
3. Launch the Summit™ Software.
4. Click on "Instrument" to start up the CyAn™ control panel.
5. Run the system clean cycle (15 min).
6. Check the lasers (488 nm and 405 nm) and let them warm up (30 min minimum for the lasers to stabilize).
7. Choose or create a database and load or design a protocol.
8. Launch the HyperCyt Summit Automator and run the "Hyperview" software.
9. Run the cytometer clean cycle with the clean and rinse solutions of your choice (EtOH, PBS, distilled water, and bleach) (*see* **Note 8**).

10. Create a new experiment file.
11. Select the wells to analyze and save your selection.
12. Determine the settings of FL1 and FL6 channels with untreated cells staining with primary and secondary antibodies. Save.
13. Start the acquisition. Protect your plate from light exposure (*see* **Note 9**).
14. Data acquisitions are automatically saved in FCS files.
15. Export your FCS files to R software and do the subsequent statistical analyses.

4 Notes

1. We have performed our statistical analysis with the R software. However, other analysis software such as FlowJo can be used.
2. The experiment is designed for human cell lines but can be adapted to murine cells.
3. The number of cells needed for the experiment depends on doubling time and vary between cell lines. Conditions might have to be adapted to the preferred cellular system. Excess cellular numbers have to be avoided for minimizing the risk of clotting. Statistical analyses are impossible if there are not enough cells.
4. You have to include at least three replicates per condition (untreated cells, positive control, negative control, drugs of interest, and untreated cells only stained with the secondary antibody) and additional wells of untreated cells to calibrate the cytometer.
5. The procedure is described for adherent cells. For nonadherent cells, do not use trypsin and directly transfer the cells into V-shaped 96-well plates.
6. DAPI is employed to evaluate cell viability and to exclude dead cells. It can be replaced by propidium iodide (PI), a fluorescent intercalating agent which cannot cross the plasma membrane of living cells. Choose between DAPI and PI according to the autofluorescence properties of your drugs.
7. Start the cytofluorometer at the beginning of the immunostaining procedure so that it is ready for use at the end of the protocol.
8. Never finish the clean cycle with PBS, which may cause the formation of precipitates.
9. Be careful: clean the CyAn™ canula with distilled water every 10 wells in order to avoid cells to accumulate inside.

Acknowledgments

GK is supported by the Ligue contre le Cancer Comité de Charente-Maritime (équipe labelisée); Agence National de la Recherche (ANR)—Projets blancs; ANR under the frame of E-Rare-2, the ERA-Net for Research on Rare Diseases; Association pour la recherche sur le cancer (ARC); Cancéropôle Ile-de-France; Chancelerie des universités de Paris (Legs Poix), Fondation pour la Recherche Médicale (FRM); a donation by Elior; the European Commission (ArtForce); the European Research Council (ERC); Fondation Carrefour; Institut National du Cancer (INCa); Inserm (HTE); Institut Universitaire de France; LeDucq Foundation; the LabEx Immuno-Oncology; the RHU Torino Lumière; the Searave Foundation; the SIRIC Stratified Oncology Cell DNA Repair and Tumor Immune Elimination (SOCRATE); the SIRIC Cancer Research and Personalized Medicine (CARPEM); and the Paris Alliance of Cancer Research Institutes (PACRI). LB is supported by Bristol Myers Squibb Foundation for Research in Immuno-Oncology (BMS).

References

1. Vitale I, Manic G, Castedo M, Kroemer G (2017) Caspase 2 in mitotic catastrophe: the terminator of aneuploid and tetraploid cells. Mol Cell Oncol 4(3):e1299274
2. Shalini S, Nikolic A, Wilson CH, Puccini J, Sladojevic N, Finnie J et al (2016) Caspase-2 deficiency accelerates chemically induced liver cancer in mice. Cell Death Differ 23 (10):1727–1736
3. Puccini J, Shalini S, Voss AK, Gatei M, Wilson CH, Hiwase DK et al (2013) Loss of caspase-2 augments lymphomagenesis and enhances genomic instability in Atm-deficient mice. Proc Natl Acad Sci U S A 110 (49):19920–19925
4. Parsons MJ, McCormick L, Janke L, Howard A, Bouchier-Hayes L, Green DR (2013) Genetic deletion of caspase-2- accelerates MMTV/c-neu-driven mammary carcinogenesis in mice. Cell Death Differ 20 (9):1174–1182
5. Dorstyn L, Puccini J, Wilson CH, Shalini S, Nicola M, Moore S et al (2012) Caspase-2- deficiency promotes aberrant DNA-damage response and genetic instability. Cell Death Differ 19(8):1288–1298
6. Andersen JL, Johnson CE, Freel CD, Parrish AB, Day JL, Buchakjian MR et al (2009) Restraint of apoptosis during mitosis through interdomain phosphorylation of caspase-2. EMBO J 28(20):3216–3227
7. Vakifahmetoglu H, Olsson M, Tamm C, Heidari N, Orrenius S, Zhivotovsky B (2008) DNA damage induces two distinct modes of cell death in ovarian carcinomas. Cell Death Differ 15(3):555–566
8. Castedo M, Perfettini JL, Roumier T, Valent A, Raslova H, Yakushijin K et al (2004) Mitotic catastrophe constitutes a special case of apoptosis whose suppression entails aneuploidy. Oncogene 23(25):4362–4370
9. Vitale I, Galluzzi L, Castedo M, Kroemer G (2011) Mitotic catastrophe: a mechanism for avoiding genomic instability. Nat Rev Mol Cell Biol 12(6):385–392
10. Castedo M, Perfettini JL, Roumier T, Andreau K, Medema R, Kroemer G (2004) Cell death by mitotic catastrophe: a molecular definition. Oncogene 23(16):2825–2837
11. Galluzzi L, Buque A, Kepp O, Zitvogel L, Kroemer G (2015) Immunological effects of conventional chemotherapy and targeted anticancer agents. Cancer Cell 28(6):690–714
12. Obeid M, Tesniere A, Panaretakis T, Tufi R, Joza N, van Endert P et al (2007) Ecto-calreticulin in immunogenic chemotherapy. Immunol Rev 220:22–34
13. Panaretakis T, Kepp O, Brockmeier U, Tesniere A, Bjorklund AC, Chapman DC et al (2009) Mechanisms of pre-apoptotic calreticulin exposure in immunogenic cell death. EMBO J 28(5):578–590

14. Apetoh L, Ghiringhelli F, Tesniere A, Obeid M, Ortiz C, Criollo A et al (2007) Toll-like receptor 4-dependent contribution of the immune system to anticancer chemotherapy and radiotherapy. Nat Med 13(9):1050–1059
15. Martins I, Wang Y, Michaud M, Ma Y, Sukkurwala AQ, Shen S et al (2014) Molecular mechanisms of ATP secretion during immunogenic cell death. Cell Death Differ 21(1):79–91
16. Sistigu A, Yamazaki T, Vacchelli E, Chaba K, Enot DP, Adam J et al (2014) Cancer cell-autonomous contribution of type I interferon signaling to the efficacy of chemotherapy. Nat Med 20(11):1301–1309
17. Michaud M, Martins I, Sukkurwala AQ, Adjemian S, Ma Y, Pellegatti P et al (2011) Autophagy-dependent anticancer immune responses induced by chemotherapeutic agents in mice. Science 334(6062):1573–1577
18. Ma Y, Adjemian S, Mattarollo SR, Yamazaki T, Aymeric L, Yang H et al (2013) Anticancer chemotherapy-induced intratumoral recruitment and differentiation of antigen-presenting cells. Immunity 38(4):729–741
19. Panaretakis T, Joza N, Modjtahedi N, Tesniere A, Vitale I, Durchschlag M et al (2008) The co-translocation of ERp57 and calreticulin determines the immunogenicity of cell death. Cell Death Differ 15(9):1499–1509
20. Madeo F, Durchschlag M, Kepp O, Panaretakis T, Zitvogel L, Frohlich KU et al (2009) Phylogenetic conservation of the preapoptotic calreticulin exposure pathway from yeast to mammals. Cell Cycle 8(4):639–642
21. Kepp O, Semeraro M, Bravo-San Pedro JM, Bloy N, Buque A, Huang X et al (2015) eIF2alpha phosphorylation as a biomarker of immunogenic cell death. Semin Cancer Biol 33:86–92
22. Bloy N, Sauvat A, Chaba K, Buque A, Humeau J, Bravo-San Pedro JM et al (2015) Morphometric analysis of immunoselection against hyperploid cancer cells. Oncotarget 6(38):41204–41215
23. Ladoire S, Senovilla L, Enot D, Ghiringhelli F, Poirier-Colame V, Chaba K et al (2016) Biomarkers of immunogenic stress in metastases from melanoma patients: correlations with the immune infiltrate. Onco Targets Ther 5(6):e1160193
24. Senovilla L, Demont Y, Humeau J, Bloy N, Kroemer G (2017) Image Cytofluorometry for the quantification of ploidy and endoplasmic reticulum stress in cancer cells. Methods Mol Biol 1524:53–64
25. Semeraro M, Adam J, Stoll G, Louvet E, Chaba K, Poirier-Colame V et al (2016) The ratio of CD8(+)/FOXP3 T lymphocytes infiltrating breast tissues predicts the relapse of ductal carcinoma in situ. Onco Targets Ther 5(10):e1218106
26. Bezu L, Sauvat A, Humeau J, Gomes-da-Silva LC, Iribarren K, Forveille S et al (2018) eIF2α phosphorylation is pathognomonic for immunogenic cell death. Cell Death and Differ 25(8):1375–1393
27. Apetoh L, Ghiringhelli F, Tesniere A, Criollo A, Ortiz C, Lidereau R et al (2007) The interaction between HMGB1 and TLR4 dictates the outcome of anticancer chemotherapy and radiotherapy. Immunol Rev 220:47–59
28. Zitvogel L, Galluzzi L, Kepp O, Smyth MJ, Kroemer G (2015) Type I interferons in anticancer immunity. Nat Rev Immunol 15(7):405–414
29. Kepp O, Senovilla L, Vitale I, Vacchelli E, Adjemian S, Agostinis P et al (2014) Consensus guidelines for the detection of immunogenic cell death. Onco Targets Ther 3(9):e955691
30. Yang H, Ma Y, Chen G, Zhou H, Yamazaki T, Klein C et al (2016) Contribution of RIP3 and MLKL to immunogenic cell death signaling in cancer chemotherapy. Onco Targets Ther 5(6):e1149673
31. Senovilla L, Vitale I, Martins I, Tailler M, Pailleret C, Michaud M et al (2012) An immunosurveillance mechanism controls cancer cell ploidy. Science 337(6102):1678–1684
32. Menger L, Vacchelli E, Adjemian S, Martins I, Ma Y, Shen S et al (2012) Cardiac glycosides exert anticancer effects by inducing immunogenic cell death. Sci Transl Med 4(143):143ra99
33. Wang W, Qin S, Zhao L (2015) Docetaxel enhances CD3+ CD56+ cytokine-induced killer cells-mediated killing through inducing tumor cells phenotype modulation. Biomed Pharmacother 69:18–23
34. Hodge JW, Garnett CT, Farsaci B, Palena C, Tsang KY, Ferrone S et al (2013) Chemotherapy-induced immunogenic modulation of tumor cells enhances killing by cytotoxic T lymphocytes and is distinct from immunogenic cell death. Int J Cancer 133(3):624–636
35. Senovilla L, Vitale I, Martins I, Kepp O, Galluzzi L, Zitvogel L et al (2013) An anticancer therapy-elicited immunosurveillance system that eliminates tetraploid cells. Onco Targets Ther 2(1):e22409
36. Castedo M, Coquelle A, Vivet S, Vitale I, Kauffmann A, Dessen P et al (2006) Apoptosis regulation in tetraploid cancer cells. EMBO J 25(11):2584–2595

Chapter 15

Quantification of eIF2α Phosphorylation Associated with Mitotic Catastrophe by Immunofluorescence Microscopy

Juliette Humeau, Lucillia Bezu, Oliver Kepp, Laura Senovilla, Peng Liu, and Guido Kroemer

Abstract

Mitotic catastrophe is an oncosuppressive mechanism that drives cells toward senescence or death when an error occurs during mitosis. Eukaryotic cells have developed adaptive signaling pathways to cope with stress. The phosphorylation on serine 51 of the eukaryotic translation initiation factor (eIF2α) is a highly conserved event in stress responses, including the one that is activated upon treatment with mitotic catastrophe inducing agents, such as microtubular poisons or actin blockers. The protocol described herein details a method to quantify the phosphorylation of eIF2α by high-throughput immunofluorescence microscopy. This method is useful to capture the 'integrated stress response', which is characterized by eIF2α phosphorylation in the context of mitotic catastrophe.

Key words Mitotic catastrophe, eIF2α, Endoplasmic reticulum, Immunofluorescence

1 Introduction

Following different kinds of stress including endoplasmic reticulum (ER) stress, hypoxia, UV irradiation, and nutrient deprivation, eukaryotic cells have developed signaling pathways to adapt to this situation or to trigger programmed cell death in case of chronic stress. The reversible phosphorylation on serine 51 of the alpha subunit of the eukaryotic translation initiation factor (eIF2α) is a highly conserved mechanism of response to stress. It can be mediated by four distinct protein kinases: HRI (heme regulated inhibitor, official name: eIF2α kinase 1, EIF2AK1) which is known for its key role during cellular development for adapting globin synthesis to the availability of heme and for promoting erythroid precursors survival when iron levels are low; PKR (protein kinase R, official name: eIF2α kinase 2, EIF2AK2) which is induced upon viral infection; PERK (PKR-like eukaryotic initiation factor 2a kinase, official name: eIF2α kinase 3, EIF2AK3) mainly activated in response to misfolded proteins in the endoplasmic reticulum lumen (unfolded protein response, UPR) and during immunogenic cell death in response to anthracycline treatment; GCN2 (general control nonderepressible 2, official name: eIF2α kinase 4, EIF2AK4) which is most known for its role in the stress response to amino acid deprivation [1, 2].

Phosphorylation of eIF2α on serine 51 (P-eIF2α) prevents the formation of ribosomal initiation complexes and leads to halt in mRNA translation. The consequent reduction in protein load within the ER lumen protects cells from ER stress-mediated apoptosis. In addition, phosphorylated eIF2α selectively favors the translation of some mRNA containing short open reading frames, like the one encoding the activating transcription factor 4 (ATF4). ATF4 is a b-ZIP transcription factor that induces growth arrest and upregulates several UPR target genes including CCAAT/enhancer-binding homologous protein (CHOP) and some chaperones [3].

Mitotic catastrophe is an oncosuppressive mechanism allowing for elimination of genomically unstable cells generated as a consequence of errors in mitosis. Mitotic failure leads either directly to mitotic catastrophe, or to tetraploid cells, which are more prone to mitotic aberrations, and thus—at the end—to mitotic catastrophe, than their diploid counterparts. Mitotic catastrophe drives the cells toward irreversible apoptosis, necrosis or senescence. Mitotic failure is induced by agents causing DNA damage (e.g., irradiation, platinum-based chemotherapeutics), by microtubular poisons including hyperpolymerizing agents (e.g., taxanes, epothilones) or depolymerizing agents (e.g., vinca alkaloids, nocodazole), by actin blockers that inhibit cytokinesis (e.g., cytochalasins) as well as by inhibitors of proteins involved in cell cycle checkpoints (e.g., auroraA/B, polo-like kinase-1, mitotic arrest deficient 2) [4–7]. It has been shown in several studies that drugs inducing mitotic failure finally cause an increase in eIF2α phosphorylation, as summarized in Table 1 [8–13].

Here, we detail a method to quantify the phosphorylation of eIF2α by high-throughput immunofluorescence microscopy using a recombinant monoclonal antibody specific for phospho-eIF2α (P-eIF2α). This protocol has been developed for human osteosarcoma U2OS cells (*see* **Note 1**) and might have to be optimized for other cell lines, especially with respect to the number of cells, the duration of treatment and the concentrations of drugs.

Table 1
Drugs that interfere with mitosis and also cause eIF2α hyperphosphorylation

Biological effect	Drug	Treatment time	Dose	Cell line	References
Inhibitor of microtubule dynamics or microtubule disassembler according to dose	Nocodazole	20 h	2 μM	HeLa	[9]
		6 h	100 nM	U2OS	[8]
		15 h	100 nM	U2OS	[10]
		48 h	100 nM	CT26	[11]
Hyperpolymerizing agents	Paclitaxel	24 h	1 μM	SK-N-SH neuroblastoma	[13]
		1 h	100 μM	U2OS	[12]
		15 h	1 μM	U2OS	[10]
		6 h	3 μM	U2OS	[8]
	Docetaxel	6 h	3 μM	U2OS	[8]
		15 h	100 nM	U2OS	[10]
	Epothilone B	15 h	30 nM	U2OS	[10]
Depolymerizing agents	Vinorelbine	1 h	25–200 μM	U2OS	[12]
		6 h	3 μM	U2OS	[8]
	Vincristine/vinblastine	1 h	100 μM	U2OS	[12]
		6 h	3 μM	U2OS	[8]
		6 h	100 nM, 1.2 μM, 3 μM	U2OS	[8]
		15 h	1 μM	U2OS	[10]
Actin blocker	Cytochalasin D	6 h	1.2 μM	U2OS	[8]
		15 h	1.2 μM	U2OS	[10]

Some agents inducing mitotic failure and mitotic catastrophe have been studied for their ability to phosphorylate eIF2α. Here, we list such agents, as well as the conditions in which they cause eIF2α phosphorylation

2 Materials

2.1 Disposables

1. 75 cm^2 polystyrene flasks for cell culture.
2. 50 mL conical centrifuge tubes.
3. 1.5 mL Eppendorf tubes.
4. Disposable pipetting reservoirs.
5. Pipette-controller and serological pipettes (5, 10, and 25 mL).
6. Micropipettes (0.2–2 μL, 1–20 μL, 20–200 μL, and 200–1000 μL) and tips.
7. Multichannel pipettes (20–200 μL).
8. Cell counting slides.
9. Plates for fluorescent microscopy (e.g., Corning™ 384 or 96-Well clear bottom black polystyrene microplates).
10. Aluminum microplate sealing tapes.

2.2 Equipment

1. Incubator allowing for standard cell culture conditions (37 °C, 5% CO_2).
2. Cell culture hood.
3. Centrifuge for 50 mL tubes.
4. Laminar flow hood (PSM).
5. Scale.
6. High-throughput automatic fluorescence imaging device with the appropriate software, for example ImageXpress micro with MetaXpress software (Molecular Device Sunnyvale, CA, USA) (*see* **Note 2**).

2.3 Reagents

1. Adequate culture medium supplemented with serum (e.g., 10% v/v fetal bovine serum) as well as 1% v/v nonessential amino acids, 1 mM sodium pyruvate, 10 mM HEPES buffer, and antibiotics (e.g., 100 U/mL penicillin G and 100 μg/mL streptomycin sulfate) if required. Store at 4 °C.
2. 0.05–0.25% trypsin solution with EDTA. Store at 4 °C.
3. 0.4% Trypan Blue solution.
4. 1× phosphate-buffered saline (PBS).
5. Hoechst 33342, trihydrochloride, 10 mg/mL solution in water (H3570, Life Technologies, Carlsbad, CA, USA). Store at 4 °C under protection from light.
6. *Fixation solution*: 4% PFA v/v, 1/5000 Hoechst v/v in PBS. Store at 4 °C under protection from light (*see* **Notes 3** and **4**).
7. *Blocking solution*: 2% BSA w/v in PBS (*see* **Note 5**).
8. Rabbit monoclonal antibody against eIF2α phosphorylated on Ser51 (ab32157, Abcam, Cambridge, UK). Store at −20 °C.
9. Alexa Fluor ™ 488 goat anti-rabbit antibody (A11034, Invitrogen, Carlsbad, CA, USA). Store at 4 °C under protection from light.

3 Method

1. After thawing, maintain the cells in culture in complete growth medium at least 1 week (*see* **Note 6**).
2. Seeding. Seed 384-well plates with 2000 cells per well in 50 μL of the appropriate culture medium (*see* **Notes 7** and **8**). Plan for at least four replicates per condition and the following controls: untreated cells, positive control(s) as well as untreated cells for staining with secondary antibody only (*see* **Notes 9–11**).
3. Treatment. Dilute the different drugs in medium in 1.5 mL Eppendorf tubes and vortex the tubes. Twenty-four hours after

seeding remove the medium from the plates and add the treatments to the appropriate wells (*see* **Note 12**). Do a stepwise treatment in order to avoid the cells from drying.

4. Fixation. After 6 h of incubation at 37 °C, 5% CO_2, remove the medium and add 30 μL/well of the fixation solution for 30–45 min at room temperature in a dark environment.

5. Discard PFA and wash with PBS (*see* **Note 13**).

6. Blocking. Add 30 μL per well of the blocking solution for 30 min to 2 h at room temperature in a dark environment.

7. Primary antibody. Discard BSA and add 20 μL of P-eIF2α rabbit antibody diluted 1/500 in the blocking solution (*see* **Note 14**). Incubate overnight at 4 °C in a dark environment.

8. After the overnight incubation with primary antibody, wash three times with PBS (*see* **Note 15**).

9. Secondary antibody. Add 30 μL of Alexa Fluor 488 goat anti rabbit diluted 1/1000 in the blocking solution. Incubate for 45 min to 2 h in a dark environment at room temperature.

10. Wash the cells three times with PBS (*see* **Note 13**).

11. Add 50 μL of PBS to each well. Avoid cells from drying by swift addition of PBS. Pipet carefully for avoiding bubbles, which is important for acquiring proper images.

12. Put the aluminum sealing tapes on the plates.

13. Image acquisition (*see* **Notes 2** and **15**). Place the plate into the high throughput automatic fluorescence imaging device in the correct orientation. By means of the software, perform acquisition using a 20× PlanApo objective. Choose a minimum of four view fields per well (*see* **Note 16**). For the DAPI channel, start with an exposure time of 30 ms and for the GFP channel of 60 ms, even though this has to be adjusted for each acquisition. Choose the optimal focus (*see* **Note 17**).

14. Image analysis (*see* **Note 2**). Perform measurements with the MetaXpress software: create a custom module that generates a mask of the nuclei based on the Hoechst 33343 fluorescence and a mask of the cytoplasm based on the GFP fluorescence (*see* **Note 18**). Measure at least the Hoechst intensity and the nuclear area in order to further exclude debris and dying cells, as well as GFP intensity in the cytoplasm. The total cellular GFP intensity and nuclear GFP intensity can also be included.

15. Data analysis (*see* **Note 19**). To analyze the data, it is recommended to export the measurements of each area of interest from MetaXpress into the R software. By means of R, it is possible to exclude debris and dying cells from further data analysis. To see if the treatment to be tested induces the

phosphorylation of eIF2α, one can either look at the average GFP intensity in the cytoplasm or consider the average of P-eIF2α positive cells based on a threshold placed between untreated and thapsigargin-treated cells.

4 Notes

1. Due to flat shape and large cytoplasm, U2OS cells allow for good cytoplasmic protein observation, making them an appropriate model for microscopy [14].

2. This protocol is adapted for image acquisition and analysis with the software MetaXpress. It has the advantage to generate files in .tiff format. However, other high throughput automatic fluorescence imaging devices can be used with their corresponding software. Alternative image analysis software can be used in subsequent steps.

3. PFA is a volatile carcinogen. It has to be manipulated under a laminar flow and disposed in an appropriate reservoir. 4% PFA solution can either be prepared from PFA powder or from a commercially available 37% PFA solution stabilized with 10% methanol. This second option is less hazardous, but the presence of methanol can cause dehydration of the tissues, which then can affect cellular morphology and induce membrane permeabilization. When using U2OS cells, it is recommended to fix the cells with PFA prepared from the 37% solution for a maximum of 1 h to avoid nuclear membrane permeabilization.

4. The fixation solution with 4% PFA and 1/5000 Hoechst 33342 is stable over time. A large volume can be prepared and stored at 4 °C for several weeks.

5. Undissolved BSA is a yellow powder that should be stored at 4 °C under protection from light. BSA requires around 15 min to be fully dissolve. The 2% BSA solution should be prepared fresh every time because it is sensitive to bacterial and fungal contaminations.

6. For maintaining of cell in culture, U2OS cells can be split at 1:6 to 1:10 ratios every 2–3 days to ensure optimal proliferation of the cells.

7. The volumes given in this protocol are for 384-well plates. This allow to minimize antibody usage, while having sufficient amount of cells for robust statistics. However, the protocol can be adapted to 96-well plates. For U2OS cells, we recommend to plate 7000 cells per well in 200 μL medium in 96-well plates.

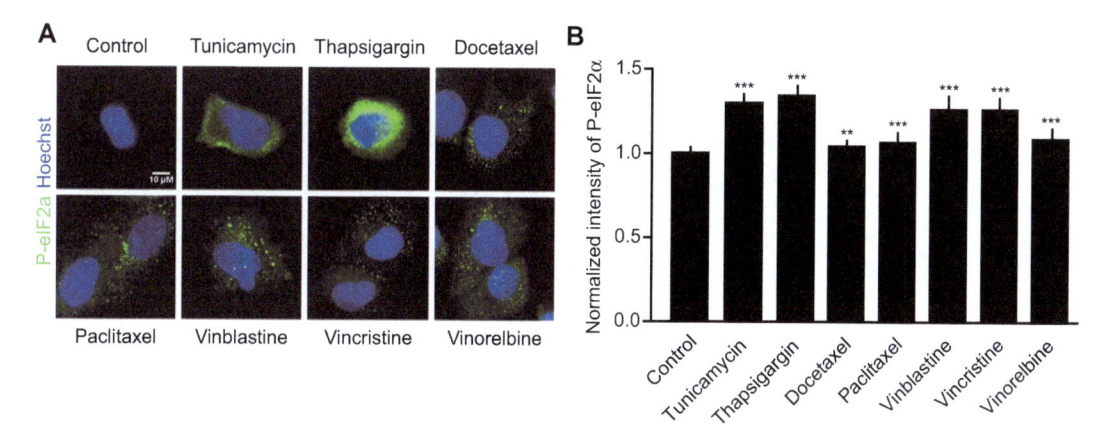

Fig. 1 Microtubules perturbing agents induce the phosphorylation of eIF2α. (**a**) U2OS cells were treated with 3 μM thapsigargin, tunicamycin, paclitaxel, docetaxel, vinorelbine, vincristine, or vinblastine for 6 h. The cells were then fixed, blocked and incubated with phosphoneoepitope-specific rabbit anti-P-eIF2α antibody overnight. Following, the cells were washed and incubated with goat anti-rabbit AF488 antibody–fluorochrome conjugate. Nuclei were stained with Hoechst 33342 and images were acquired using a 20× PlanApo objective. (**b**) The experiment was performed four times in quadruplicates. In each experiment, the ratio of cytoplasmic GFP intensity was measured and the ratio of treated to untreated control is depicted. Mean ± SEM of four independent experiments is shown. A statistical analysis with a Student's *t*-test was performed (*$p < 0.05$ **$p < 0.01$ ***$p < 0.001$)

8. The number of cells per well should reach a maximum of 60–80% of confluency. This is crucial since the phosphorylation of eIF2α is highly sensitive to confluence.

9. Do not use border wells to avoid border effects.

10. Use one or two positive controls for ER stress. Tunicamycin and thapsigargin are efficient inducers of eIF2α phosphorylations. For U2OS cells, we recommend to use 1–10 μM tunicamycin and 0.5–10 μM thapsigargin (Fig. 1). However, other drugs can be used as positive controls for eIF2α phosphorylation as well (*see* Table 2).

11. Always plan a control with the secondary antibody alone to exclude nonspecific binding. If nonspecific binding occurs, this must be taken into account as a baseline value when analyzing data.

12. When the cells are treated, fresh medium must be changed in control conditions for avoiding false-positive results due to the overconsumption of nutrients.

13. Each washing step consists in flipping the plate over a beaker to remove the solution, adding 50 μL/well PBS and emptying the plate again.

14. In this protocol, the antibody is small enough to cross the plasma membrane without the need of prior detergent-mediated permeabilization.

Table 2
Examples of agents inducing eIF2α phosphorylation

Name	Mechanism of action
Activators of ER stress	
Tunicamycin	Inhibitor of N-acetylglucosamine phosphotransferase leading to N-glycosylation of newly synthetized proteins inhibition
Thapsigargin	Inhibitor of sarco/endoplasmic calcium ATPase (SERCA) leading to depletion of the ER calcium stores
Brefeldin A	Inhibitor of protein transport from the endoplasmic reticulum to the Golgi by preventing formation of COPI-mediated transport vesicles
DL-Dithiothreitol	Reducing agent by blocking disulfide bonds formation
Eeyarestatin I	Inhibitor of ER-associated proteins translocation and degradation
Inhibitor of eIF2a phosphatases	
Salubrinal	Inhibitor of the PP1/GADD34 complex
Tautomycin	Inhibitor of PP1 and PP2A
Guanabenz	Inhibitor of GADD34
Sephin 1	Inhibitor of GADD34
Calyculin A	Inhibitor of PP1 and PP2A

This table contains a nonexhaustive list of agents inducing eIF2α phosphorylation. They are listed according to their mode of action: interference with ER function or inhibition eIF2α phosphatases

15. Before image acquisition, keep the plate at room temperature for 30 min to avoid condensation.

16. We recommend taking at least four pictures per well to acquire information on a sufficient amount of cells and to choose sites which are not too close to the border of the wells to avoid side effects arising from uneven plate bottoms. If the treatment is toxic, number of pictures per well to gain information on a minimum number of cells might have to be increased. However, acquiring too many pictures per well may increase the acquisition time and hence slows down the process.

17. When imaging the plate, the best focus might not be the same for each site since the plates have certain variations on bottom thickness. In this case, a z-stack can be acquired and subsequently collapsed to allow for adequate focusing on each site. The number of z-plans for each site and the interval in between has to be determined for each acquisition according to the plastic and biological material used.

18. When detecting cytoplasmic masks, some cells with inherently low cytoplasmic background might be neglected. Therefore,

we recommend to include a step of image analysis that allows to define a ring around the nucleus that can be defined as cytoplasm in cells where no cytoplasm was detected.

19. R is well suited for treating data from images analysis. However, it is also possible to directly export the images measurements from MetaXpress into Excel or to use other data analysis software.

Acknowledgments

GK is supported by the Ligue contre le Cancer Comité de Charente-Maritime (équipe labelisée); Agence National de la Recherche (ANR)—Projets blancs; ANR under the frame of E-Rare-2, the ERA-Net for Research on Rare Diseases; Association pour la recherche sur le cancer (ARC); Cancéropôle Ile-de-France; Chancelerie des universités de Paris (Legs Poix), Fondation pour la Recherche Médicale (FRM); a donation by Elior; the European Commission (ArtForce); the European Research Council (ERC); Fondation Carrefour; Institut National du Cancer (INCa); Inserm (HTE); Institut Universitaire de France; LeDucq Foundation; the LabEx Immuno-Oncology; the RHU Torino Lumière; the Searave Foundation; the SIRIC Stratified Oncology Cell DNA Repair and Tumor Immune Elimination (SOCRATE); the SIRIC Cancer Research and Personalized Medicine (CARPEM); and the Paris Alliance of Cancer Research Institutes (PACRI). JH is supported by the Fondation Philanthropia. LB is supported by Bristol Myers Squibb Foundation for Research in Immuno-Oncology (BMS). PL is supported by the China Scholarship Council.

References

1. Donnelly N, Gorman AM, Gupta S, Samali A (2013) The eIF2alpha kinases: their structures and functions. Cell Mol Life Sci 70(19):3493–3511. https://doi.org/10.1007/s00018-012-1252-6

2. Panaretakis T, Kepp O, Brockmeier U, Tesniere A, Bjorklund AC, Chapman DC, Durchschlag M, Joza N, Pierron G, van Endert P, Yuan J, Zitvogel L, Madeo F, Williams DB, Kroemer G (2009) Mechanisms of pre-apoptotic calreticulin exposure in immunogenic cell death. EMBO J 28(5):578–590. https://doi.org/10.1038/emboj.2009.1

3. Walter P, Ron D (2011) The unfolded protein response: from stress pathway to homeostatic regulation. Science 334(6059):1081–1086. https://doi.org/10.1126/science.1209038

4. Vitale I, Galluzzi L, Castedo M, Kroemer G (2011) Mitotic catastrophe: a mechanism for avoiding genomic instability. Nat Rev Mol Cell Biol 12(6):385–392. https://doi.org/10.1038/nrm3115

5. Denisenko TV, Sorokina IV, Gogvadze V, Zhivotovsky B (2016) Mitotic catastrophe and cancer drug resistance: a link that must to be broken. Drug Resist Updat 24:1–12. https://doi.org/10.1016/j.drup.2015.11.002

6. Portugal J, Mansilla S, Bataller M (2010) Mechanisms of drug-induced mitotic catastrophe in cancer cells. Curr Pharm Des 16(1):69–78

7. Ganguly A, Yang H, Sharma R, Patel KD, Cabral F (2012) The role of microtubules and their dynamics in cell migration. J Biol Chem 287

(52):43359–43369. https://doi.org/10.1074/jbc.M112.423905

8. Bezu L, Sauvat A, Humeau J, Gomes-da-Silva LC, Iribarren K, Forveille S, Garcia P, Zhao L, Liu P, Senovilla L, Kepp O, Kroemer G (2018) eIF2α phosphorylation is pathognomonic for immunogenic cell death. Cell Death Differ 25(8):1375–1393

9. Coldwell MJ, Cowan JL, Vlasak M, Mead A, Willett M, Perry LS, Morley SJ (2013) Phosphorylation of eIF4GII and 4E-BP1 in response to nocodazole treatment: a reappraisal of translation initiation during mitosis. Cell Cycle 12(23):3615–3628. https://doi.org/10.4161/cc.26588

10. Senovilla L, Vitale I, Martins I, Tailler M, Pailleret C, Michaud M, Galluzzi L, Adjemian S, Kepp O, Niso-Santano M, Shen S, Marino G, Criollo A, Boileve A, Job B, Ladoire S, Ghiringhelli F, Sistigu A, Yamazaki T, Rello-Varona S, Locher C, Poirier-Colame V, Talbot M, Valent A, Berardinelli F, Antoccia A, Ciccosanti F, Fimia GM, Piacentini M, Fueyo A, Messina NL, Li M, Chan CJ, Sigl V, Pourcher G, Ruckenstuhl C, Carmona-Gutierrez D, Lazar V, Penninger JM, Madeo F, Lopez-Otin C, Smyth MJ, Zitvogel L, Castedo M, Kroemer G (2012) An immunosurveillance mechanism controls cancer cell ploidy. Science 337(6102):1678–1684. https://doi.org/10.1126/science.1224922

11. Senovilla L, Demont Y, Humeau J, Bloy N, Kroemer G (2017) Image cytofluorometry for the quantification of ploidy and endoplasmic reticulum stress in cancer cells. Methods Mol Biol 1524:53–64. https://doi.org/10.1007/978-1-4939-6603-5_3

12. Szaflarski W, Fay MM, Kedersha N, Zabel M, Anderson P, Ivanov P (2016) Vinca alkaloid drugs promote stress-induced translational repression and stress granule formation. Oncotarget 7(21):30307–30322. https://doi.org/10.18632/oncotarget.8728

13. Tanimukai H, Kanayama D, Omi T, Takeda M, Kudo T (2013) Paclitaxel induces neurotoxicity through endoplasmic reticulum stress. Biochem Biophys Res Commun 437(1):151–155. https://doi.org/10.1016/j.bbrc.2013.06.057

14. Perlman ZE, Slack MD, Feng Y, Mitchison TJ, Wu LF, Altschuler SJ (2004) Multidimensional drug profiling by automated microscopy. Science 306(5699):1194–1198

Chapter 16

Clonogenic Assays to Detect Cell Fate in Mitotic Catastrophe

José Manuel Bravo-San Pedro, Oliver Kepp, Allan Sauvat, Santiago Rello-Varona, Guido Kroemer, and Laura Senovilla

Abstract

Mitotic catastrophe (MC) is a cell death modality induced by DNA damage that involves the activation of cell cycle checkpoints such as the "DNA structure checkpoint" and "spindle assembly checkpoint" (SAC) leading to aberrant mitosis. Depending on the signal, MC can drive the cell to death or to senescence. The suppression of MC favors aneuploidy. Several cancer therapies, included microtubular poisons and radiations, trigger MC. The clonogenic assay has been used to study the capacity of single cells to proliferate and to generate macroscopic colonies and to evaluate the efficacy of anticancer drugs. Nevertheless, this method cannot analyze MC events. Here, we report an improved technique based on the use of human colon cancer HCT116 stable expressing histone H2B-GFP and DsRed-centrin proteins, allowing to determine the capacity of cells to proliferate, and to determine changes in the nucleus and centrosomes.

Key words Mitotic catastrophe, Clonogenic assay, Histone, Centrin

1 Introduction

DNA damage can arise from endogenous and environmental sources. Endogenous damage is caused by cellular metabolic products [1] or by reactive oxygen species [2] leading to oxidation, alkylation, hydrolysis, and mismatch of bases as well as bulky adduct formation. Exogenous sources of DNA damage include UV [3], X-, [4] and, ionizing radiations [5, 6], elevated temperature [7], industrial chemicals [8], drugs [9], plant toxins [10, 11], and viruses [12, 13]. DNA damage results in double- and single-strand breaks (DSBs and SSBs, respectively), as well as damage and loss of bases. When DNA damage occurs, cells either survival or die. To survive, cells arrest their cell cycle because the DNA must be repaired before genomic information is duplicated and hence transmitted to the next generation. Cells have developed multiple mechanisms of DNA damage response (DDR) including base excision repair (BER), mismatch repair (MMR), nucleotide excision repair (NER), cross-link repair (ICL), as well as double-strand break (DSB) repair pathways. These repair mechanisms are regulated by different phosphatidylinositol 3-kinase-related kinases (PIKK) including DNA-dependent protein kinase (DNA-PKcs), ataxia telangiectasia-mutated (ATM), and ATM and Rad3-related (ATR) [14–16]. Failure to repair DNA leads to senescence, apoptosis, necrosis, or mitotic catastrophe (MC) [15].

MC is a cell death modality induced by DNA damage that has been first described in temperature-sensitive lethal phenotypes of the fission yeast *Schizosaccharomyces pombe* involving the mitosis regulators cyclin-dependent kinase 1 (Cdk1), M-phase inducer phosphatase (Cdc25) and mitosis inhibitor protein kinase wee1 (Wee1). Abnormalities in the expression of these proteins lead to aberrant mitosis mainly due to chromosome missegregation [17–19]. The regulation of mitotic entry by these proteins and others is conserved from yeast to humans [20]. At present, it is widely accepted that MC implies proteins forming the "DNA structure checkpoint" which arrests cells at the G2/M transition, and the "spindle assembly checkpoint" (SAC) which prevents anaphase until all chromosomes have obtained bipolar attachment, such as the afore mentioned ATM, ATR, as well as checkpoint kinase (Chk) 1, Chk2, and polo-like kinase (Plk)1, among others. Dysfunction of these checkpoints can cause premature mitosis before the completion of DNA repair. Importantly, there is no MC without premature entry into mitosis. As a consequence of the alterations in G2/M transition checkpoint and SAC, MC can result micronucleation (delayed chromosomes that during karyokinesis fail to join one of the main nuclei and instead recruit their own nuclear envelope components), multinucleation (two or more nuclei contained in the same cell due to cytokinesis failure), overduplication of centrosomes causing multipolar mitosis, or failure of centrosomes to undergo duplication [15, 21–23]. Depending on the DNA damage signal, MC ends in cell death through apoptosis or necrosis or results in senescence, i.e., an irreversible cell cycle arrest [21, 24, 25]. Failure to undergo MC may result into the formation of polyploid cells that then undergo multipolar divisions, leading to the generation of aneuploid cells [26]. Of note, several cancer therapies based on microtubular poisons (such as taxanes, epothilones, or vinca alkaloids trigger mitotic catastrophe in tumor cells [14, 15, 23].

In 1956, Puck and Marcus developed a method, named clonogenic cell survival assay or clonogenic assay, to study, for the first time, an X-radiation dose–response curve in mammalian cells [27]. The clonogenic assay allows to study the capacity of single cells to proliferate and to generate macroscopic colonies, usually after treatment of the cells with different anticancer agents. Such treatments include not only radiation but also other cancer therapies such as MC inducers, either alone or in combination [28, 29], and the decrease of the number and size of colonies compared to untreated controls yield information on the potency of anticancer agents. However, it is not possible to quantify MC induced by such treatments. Here, we report a refinement of the classical clonogenic assay based on the use of human colon carcinoma cell line, HCT116 (which has proved to be a good MC model) [30], that has been engineered to stably express a histone H2B-green fluorescent protein (H2B-GFP) and a *Discosoma striata* red fluorescent-centrin protein (DsRed-Centrin chimera) [31]. By combining high-resolution fluorescence microscopy and automated image analysis, we do not only measure the number of proliferative colonies but also assess nuclear changes and alterations in centrosome number that may accompany MC.

2 Materials

2.1 Disposables

1. 75 cm^2 flasks for cell culture (Corning, New York, NY, USA).
2. KOVA™ Glasstic™ slides 10 with grids (Fisher Scientific, Hampton, NH, USA).
3. Flat bottom cell culture plates 6-well (Corning, New York, NY, USA).
4. Falcon® 96-well black with clear flat bottom TC-treated imaging microplates (BD Biosciences, San Jose, CA, USA).
5. Aluminum microplate sealing tapes.

2.2 Equipment

1. ImageXpress Micro XLS Widefield High-Content Analysis System (Molecular Device LLC, Sunnyvale, CA, USA) equipped with a Sola light source, standard DAPI, GFP, Texas Red, Cy3, and Cy5 filters sets as well as 10×, 20×, and PlanFluor and PlanApo objectives.

2.3 Reagents

2.3.1 Culture

1. Complete growth medium for HCT116 cells. McCoy's medium (Gibco, Gaithersburg, MD, USA) supplemented with 1 mM sodium pyruvate, 100 mM HEPES buffer, 100 U/mL penicillin G, 100 μg/mL streptomycin sulfate, and 10% fetal bovine serum (FBS).
2. TrypLE™ Express Enzyme with EDTA without red phenol (Gibco), stable for 24 months stored at 4 °C (*see* **Note 1**).
3. Trypan Blue Solution (Sigma-Aldrich, St Louis, MO, USA) stored at room temperature (*see* **Notes 2 and 3**).

2.3.2 Treatments with Mitotic Catastrophe and Apoptotic Inducers

1. Eg5 Inhibitor III, Dimethylenastron (DIMEN, Merck KGaA (Calbiochem brand), Darmstadt, Germany), stock solution in DMSO, 10 mM, stored at −20 °C (*see* **Notes 4 and 5**).
2. Nocodazole (NOC, Sigma-Aldrich, St Louis, MO, USA), stock solution in DMSO, 1 mM, stored at −20 °C (*see* **Notes 6 and 7**).

3. Paclitaxel (PTX, Sigma-Aldrich, St Louis, MO, USA), stock solution in DMSO, 3 mM, stored at −20 °C (*see* **Notes 8** and **9**).

4. 7-hydroxystaurosporine (UCN-01, Sigma-Aldrich, St Louis, MO, USA), stock solution in DMSO, 10 mM, stored at −20 °C (*see* **Note 10**).

2.3.3 Clonogenic Assays

1. Crystal Violet solution (Sigma-Aldrich, St Louis, MO, USA), stored at room temperature (*see* **Notes 11** and **12**).

2. Formalin solution (Sigma-Aldrich, St Louis, MO, USA), stock solution at 10% in distilled or deionized water (*see* **Note 13**).

3. PBS pH 7.4 (Gibco), stored at 4 °C.

4. Deionized water.

3 Methods

3.1 Cell Culture

1. HCT116 H2B-GFP DsRed-Centrin cells are maintained in complete growth medium in 75 cm^2 flasks (37 °C, 5% CO_2) and split at 1:5.

2. When confluence approaches 70–80%, cells are washed with PBS pH 7.4 and detached with TrypLE™ enzyme (*see* **Note 14**) to constitute fresh maintenance cultures and to provide cells for experimental determinations.

3.2 Classical Clonogenic Assay

1. *Seeding*: For classical clonogenic assay experiments, HCT116 H2B-GFP DsRed-Centrin cells are seeded (500 cells/well) in 6-well plates (*see* **Note 15**).

2. *Treatment*: 24 h after seeding, cells are treated with various drug concentrations for 48 h. Then cells are washed to remove the drugs and cultured in normal cultured medium for 6 more days (*see* **Note 16**).

3. *Staining*: At the end point, cells are stained with Crystal Violet. Culture medium is removed from the wells and 500 μL of Crystal Violet is added. After 10 min, the Crystal Violet solution is discarded, and wells are washed with deionized water until the wells are clean enough to count the colonies.

4. *Quantification*: Once the plates are dry, the colonies can be quantified, either in terms of the number of colonies or colony area through the homonym Image J plugin [32] (Fig. 1).

3.3 Fluorescent Clonogenic Assay Microscopy

1. *Seeding*: HCT116 H2B-GFP DsRed-Centrin cells are seeded (100 cells/well) into TC-treated Falcon™ 96 well imaging microplates (black with clear flat bottoms).

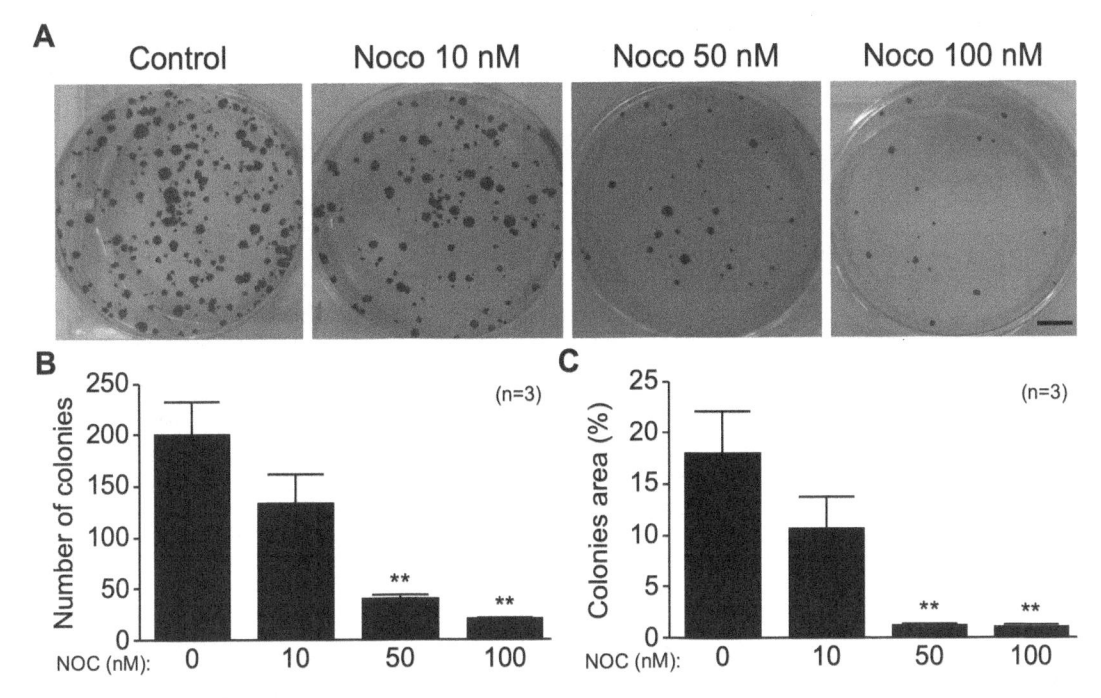

Fig. 1 Classical clonogenic assay. Untreated (0) and nocodazole (NOC)-treated HCT116 H2B-GFP DsRed-Centrin cells were stained with Crystal Violet at end point to detect colonies. Four images representative for each gradual concentration (0, 25, 50, and 100 nM NOC) are shown (**a**). Quantification of colonies number by hand (**b**) or colony area by Image J (**c**) are the two parameters measured with the classical clonogenic assay. Error bars indicate SEM. One-tailed Student's t tests were used for statistical comparisons. **$p < 0.01$, as compared to untreated cells. NOC, nocodazole. Calibration bar: 5 mm. Note that nocodazole reduces the capacity of cells to proliferate

2. *Treatment*: 24 h after seeding, cells are treated with MC inducers at low or high concentrations: DIMEN (100 and 500 nM), NOC (10 and 50 nM), PXT (10 and 50 nM), and UCN-01 (50 and 250 nM) for 48 h. Then cells are washed to remove the MC inducer and cultured in normal cultured medium for 4 more days.

3. *Fixation*: At the end point, cells are washed twice with PBS pH 7.4, and fixed for 20 min in cold 4% (v/v) formalin solution. Then cells are washed again with PBS pH 7.4 (*see* **Note 17**).

4. Microcopic detection of fluorescent colonies. Plates are imaged on the ImageXpress Micro XLS Widefield High-Content Analysis System using an ×10 Plan Fluor objective (Nikon, Tokyo, Japan) and a 16-bit monochromes sCMOS PCO.edge 5.5 camera (PCO Kelheim, Germany). Images are acquired to obtain a 95% field-of-view. Five × 5 adjacent images need to be acquired to cover an entire well of a 96 well-plate, and GFP fluorescence is acquired using a standard GFP bandpass filter set (Ex 472/30–25 and Em 520/35–25 nm from Semrock, Rochester, NY, US), and 75 ms of exposure (Fig. 2) (*see* **Note 18**).

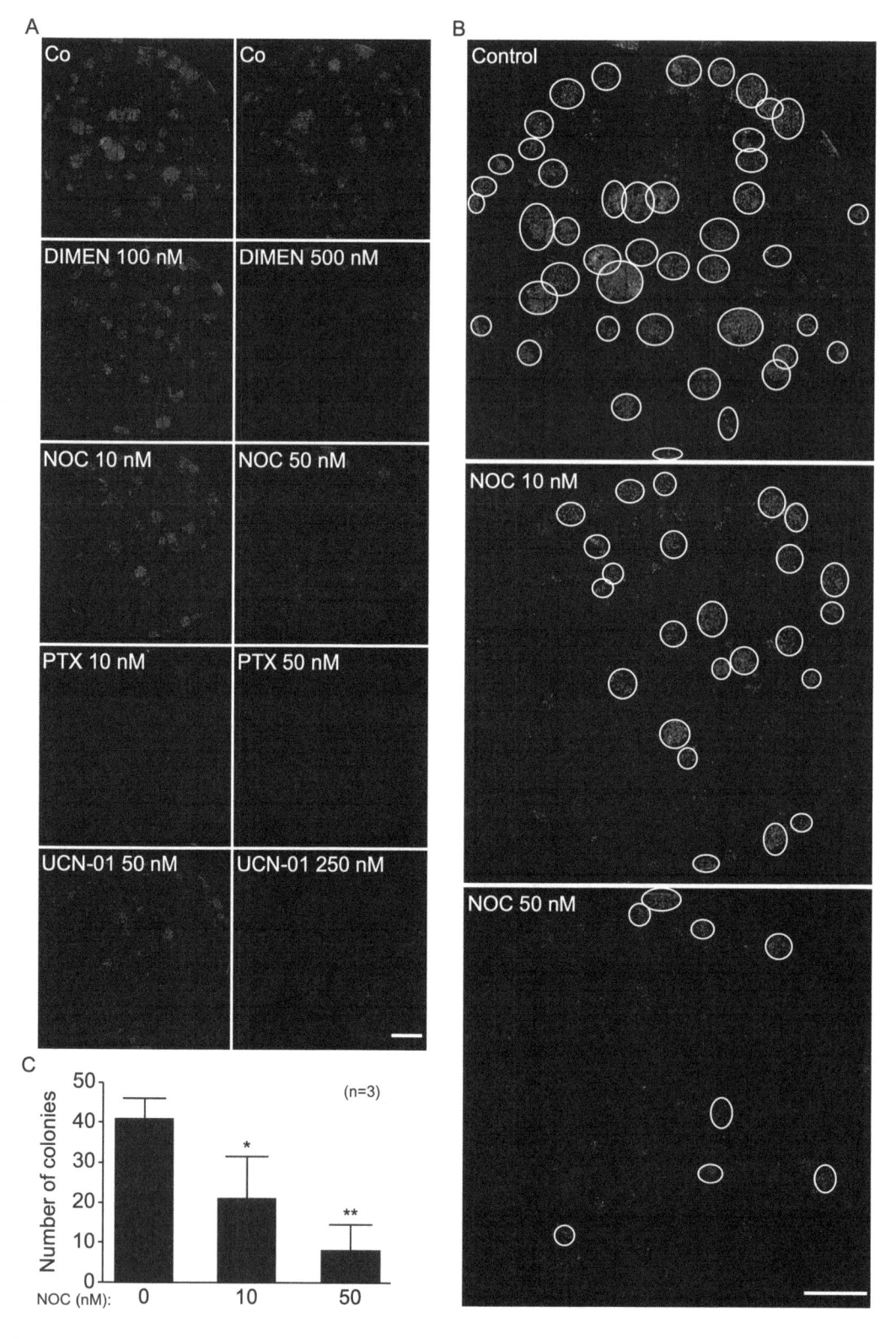

Fig. 2 Fluorescent clonogenic assay microscopy. HCT116 H2B-GFP DsRed-Centrin cells are treated for 48 h with different MC or apoptosis-inducing agents at different concentrations: DIMEN 100 and 500 nM, NOC

3.4 Mitotic Catastrophe Analysis

1. The same plates used as above (GFP acquisition at a magnification of 10×) are used for the fine analysis of MC at higher magnification.

2. Plates are imaged on the ImageXpress Micro XLS Widefield High-Content Analysis System using an ×40 Plan Apo objective (Nikon, Tokyo, Japan) and a 16-bit monochromes sCMOS PCO.edge 5.5 camera (PCO). Images are acquired to cover 95% field-of-view, meaning that 10 × 10 adjacent images per well are taken (see **Note 19**). GFP fluorescence is acquired (GFP bandpass filter set Xx 472/30–25 and Em 520/35–25 nm, 75 ms exposure) for nuclear detection. For centrosome detection, DsRed fluorescence is acquired (Texas Red bandpass filter set (Ex 562/40–25 and Em 624/40–25 nm from Semrock, 400 ms exposure) using a laser-based autofocus and z-stack acquisitions, consisting of 8 planes with 2 μm distance for each image to balance plate bottom variations, are taken. Stacks need to be collapsed using best focus and maximum projection algorithms for the detection of nuclei and centrosomes, respectively (see **Notes 20** and **21**) (Fig. 3).

3. Image analysis. Images are analyzed by means of the MetaXpress software (Molecular Devices, Sunnyvale, CA, US) using an algorithm created with the custom module editor (CME, Molecular Devices) that allows to segment each single cell into cytoplasm and nucleus and to simultaneously detect centrin spots. The following parameters are measured: number of cells (mask: cells = nuclei); mitotic cells (two parameters are used: nuclear granularity, as the ratio between the average difference in fluorescence of adjacent pixels within one nucleus and its overall average fluorescence, and nuclear densitometry, as ratio between the mean fluorescence intensity and the nuclear area) (see **Note 22**); polyploidy (mask: nuclear area) (see **Note 23**); and the number of centrosomes (mask: centrin spots) (see **Note 24**) (Fig. 4).

Fig. 2 (continued) 10 and 50 nM, PXT 10 and 50 nM, and UCN-01 50 and 250 nM. The cells were washed and cultured in normal culture medium for 4 days. After fixation, GFP fluorescence was acquired using 10× Plan Fluor objective to show the colonies. Qualitative (**a**) analysis and quantitative (**b** and **c**) analysis were carried out. Representative images are shown in (**a** and **b**). Calibration bar: 1 mm. The number of colonies detected in (**b**) is plotted in (**c**). Error bars indicate SEM. One-tailed Student's *t* tests were used for statistical comparisons. *$p < 0.05$, **$p < 0.01$, as compared to untreated cells. *DIMEN* dimethylenastron, *NOC* nocodazole, *PXT* paclitaxel, *UCN-01* 7-hydroxystaurosporine

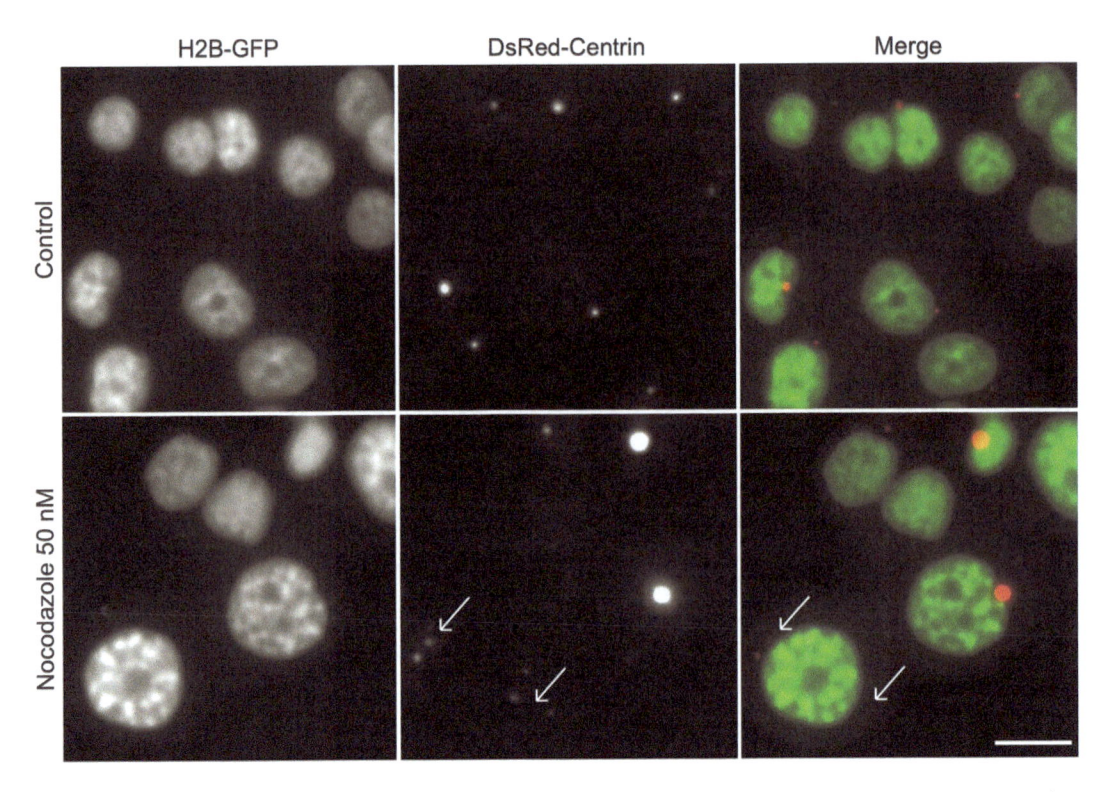

Fig. 3 Nuclei and centrin detection. Images from HCT116 H2B-GFP DsRed-Centrin cells that were untreated or treated with NOC 50 nM were acquired using a 40× Plan Apo objective to detect nuclei (H2B-GFP) and centrosomes (DsRed-centrin). Representative images of control and nocodazole-treated cells are shown. Arrows point to supernumerary centrosomes. Calibration bar: 10 μm. Note that nocodazole-treated cells exhibit an increase in nuclear size and in supernumerary centrosomes. *NOC* nocodazole

4 Notes

1. TrypLE™ contains no substances that are considered to be hazardous. Nevertheless, it may be harmful if inhaled or swallowed, and it may cause eye and skin irritation.

2. Trypan Blue is a dye that is not absorbed by healthy viable cells, but stains cells with a damaged cell membrane, hence allowing to distinguish live and dead cells. It is used at 10% (v/v) in PBS.

3. Trypan Blue may cause cancer. Always handle Trypan Blue using protective eye shields, gloves and respiratory filters.

4. DIMEN is an inhibitor of the microtubule-stimulated ATPase activity of the mitotic kinesin Eg5 (also known as kinesin spindle protein, KSP) [33].

5. Undissolved DIMEN appears as a light-yellow powder and may be stored at 4 °C under protection from light. After reconstitution, aliquots are stable for 6 months at −20 °C.

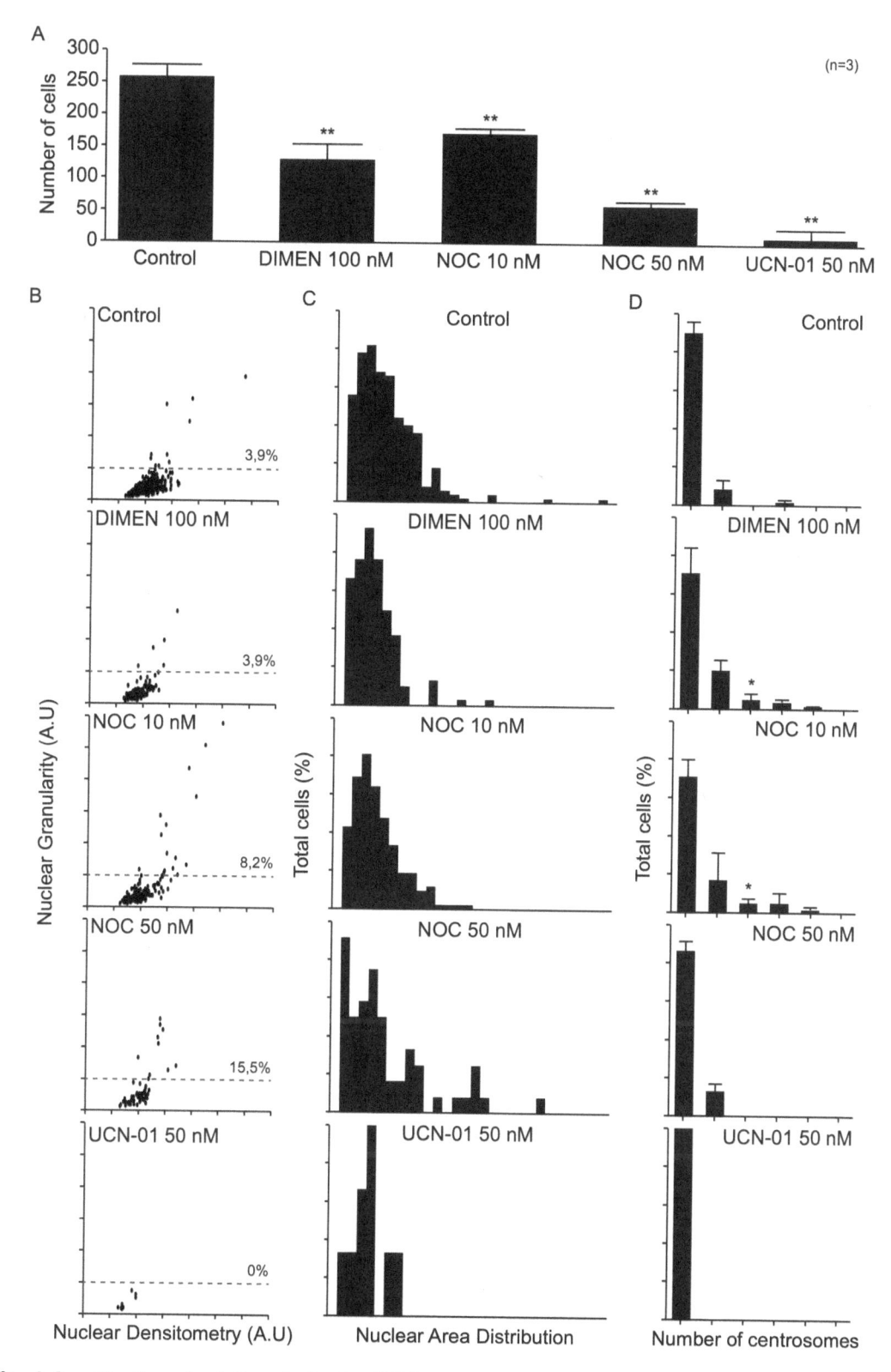

Fig. 4 Quantification of mitotic catastrophe. HCT116 H2B-GFP DsRed-Centrin cells were cultured in the absence or presence of DIMEN 100 nM, NOC 10 nM, NOC 50 nM, or UCN-01 50 nM. Images were acquired using a 40× Plan Apo objective and analyzed by means of MetaXpress software for the number of cells (**a**), mitotic cells (**b**), ploidy (**c**), and number of centrosomes (**d**). The red line in **b** indicates the threshold to

6. NOC interferes with the microtubules polymerization, thus blocking vesicular transport, mitotic spindle formation, and cytokinesis [34].

7. NOC is hazardous to health and can be fatal if inhaled, swallowed or absorbed through skin. Always handle NOC using protective eye shields, gloves, and respiratory filters. Undissolved NOC appears as a white powder and may be stored at 4 °C under protection from light for 2–3 years.

8. PTX targets tubulin causing defects in mitotic spindle assembly, chromosome segregation, and cell division [35, 36].

9. PTX is hazardous to health if inhaled, swallowed, or in case of skin or eye contact. Always handle PTX using safety glasses, gloves, and respiratory filters. Undissolved PTX may be stored at 4 °C under light protection.

10. UCN-01 is a staurosporine (STS) derivative, a tyrosine kinase inhibitor that induces apoptosis but not MC [31, 37].

11. Crystal Violet is the dye classically used in Gram staining for classifying bacteria but nowadays is also used for rendering mammalian cells visible [38].

12. Crystal Violet is hazardous to health and can be fatal if swallowed or absorbed through skin.

13. Formalin solution contains approximately 4% of formaldehyde, is hazardous to health and can be fatal if swallowed or absorbed through skin.

14. Before trypsinization, cells grown to approximately 80% of confluency are washed with 10 mL of PBS pH 7.4. Then, 4 mL TrypLE™ Express is added for 75 cm^2 flasks and flasks are placed for 5 min at 37 °C to completely detach the cells.

15. After trypsinization, cells are washed with fresh medium and counted using KOVA™ Glasstic™ slides 10 with grids, while excluding dead cells by staining them with Trypan Blue (10% in PBS).

16. After 48 h of treatment cells are mostly arrested in their cell cycle. However, at later time points cells may escape from this cell cycle arrest or manifest signs of MC.

Fig. 4 (continued) distinguish mitotic cells (top half), corresponding to 1000 arbitrary units (A.U.). The scale (in **c** and **d**) ranges from 0% to 100% of total cells. On to six centrosomes were found per cell (**d**). Error bars indicate SEM. One-tailed Student's *t* test was used for statistical comparisons. *$p < 0.05$, **$p < 0.01$, as compared to untreated cells. Note that all mitotic catastrophe and apoptotic agents have a cytotoxic effect and that only morphological analyses can distinguish between mitotic catastrophe mitotic and apoptosis. DIMEN and NOC increase the percentage of mitotic cells as well as the number of centrosomes. NOC increases cell ploidy (*see* **Note 25**). *DIMEN* dimethylenastron, *NOC* nocodazole, *UCN-01* 7-hydroxystaurosporine

17. Once the fixation is done, plates can be stored at 4 °C until they are used.
18. Colonies are quantified by eye (Fig. 2b) and data are processed using Excel software (example of quantification Fig. 2b, c).
19. Using 10 × 10 multisite adjacent images, just the center of the well will be acquired.
20. Centrin protein is found in the centrosomes.
21. Centrosomes are not necessarily at the same focus than nuclei. To detect the maximum number of centrosomes, it is necessary to carry out a z-stack acquisition for DsRed fluorescence.
22. Granularity is related to DNA condensation, which increases in mitotic or apoptotic cells. Mitotic arrest is accompanied by increased granularity due to chromatin condensation, accompanied by increased mean density.
23. Nuclear area is an indirect measure of ploidy status [39–41].
24. Automatic detection of the number of centrosomes per cell is calculated in a region determined by the nuclear area (GFP signal) plus a perinuclear extension zone of 10 pixels. The number of centrosomes within this area is determined using the DsRed-Centrin signal.
25. Treatment with MC-inducing agents during 48 h triggers an arrest of the cell cycle. Once the MC inducer has been washed out and the cells are placed in normal culture media, a variable portion of cells gradually recovers from the cell cycle arrest and undergoes MC or escapes from MC. Cell cycle divisions after mitotic catastrophe can lead to multipolar asymmetric divisions resulting in the generation of aneuploid cells [22].

Acknowledgments

GK is supported by Comité de Charente-Maritime (équipe labelisée); Agence National de la Recherche (ANR)—Projets blancs; ANR under the frame of E-Rare-2, the ERA-Net for Research on Rare Diseases; Association pour la recherche sur le cancer (ARC); Cancéropôle Ile-de-France; Chancelerie des universités de Paris (Legs Poix), Fondation pour la Recherche Médicale (FRM); a donation by Elior; the European Commission (ArtForce); the European Research Council (ERC); Fondation Carrefour; Institut National du Cancer (INCa); Inserm (HTE); Institut Universitaire de France; LeDucq Foundation; the LabEx Immuno-Oncology; the RHU Torino Lumière; the Seerave Foundation; the SIRIC Stratified Oncology Cell DNA Repair and Tumor Immune Elimination (SOCRATE); the SIRIC Cancer Research and Personalized Medicine (CARPEM); and the Paris Alliance of Cancer Research

Institutes (PACRI). LS is supported by the Institut national de la santé et de la recherche médicale (INSERM) and the Association pour la recherche sur le cancer (ARC) (PJA 20151203519). JMB-SP is currently funded by "Ramon y Cajal Program" (RYC-2018-025099-I) and supported by Spain's Ministerio de Ciencia e Innovacion (PID2019-108827RA-I00).

References

1. Burcham PC (1999) Internal hazards: baseline DNA damage by endogenous products of normal metabolism. Mutat Res 443(1–2):11–36

2. Moloney JN, Cotter TG (2018) ROS signalling in the biology of cancer. Semin Cell Dev Biol 80:50–64

3. Schuch AP, Moreno NC, Schuch NJ, Menck CFM, Garcia CCM (2017) Sunlight damage to cellular DNA: focus on oxidatively generated lesions. Free Radic Biol Med 107:110–124

4. Parplys AC, Petermann E, Petersen C, Dikomey E, Borgmann K (2012) DNA damage by X-rays and their impact on replication processes. Radiother Oncol 102(3):466–471

5. Tusher VG, Tibshirani R, Chu G (2001) Significance analysis of microarrays applied to the ionizing radiation response. Proc Natl Acad Sci U S A 98(9):5116–5121

6. Sutherland BM, Bennett PV, Sidorkina O, Laval J (2000) Clustered DNA damages induced in isolated DNA and in human cells by low doses of ionizing radiation. Proc Natl Acad Sci U S A 97(1):103–108

7. Kantidze OL, Velichko AK, Luzhin AV, Razin SV (2016) Heat stress-induced DNA damage. Acta Nat 8(2):75–78

8. Kailasam S, Rogers KR (2007) A fluorescence-based screening assay for DNA damage induced by genotoxic industrial chemicals. Chemosphere 66(1):165–171

9. Muller S (2017) DNA damage-inducing compounds: unraveling their pleiotropic effects using high throughput sequencing. Curr Med Chem 24(15):1558–1585

10. Fadlalla K, Watson A, Yehualaeshet T, Turner T, Samuel T (2011) *Ruta graveolens* extract induces DNA damage pathways and blocks Akt activation to inhibit cancer cell proliferation and survival. Anticancer Res 31(1):233–241

11. Lam M, Carmichael AR, Griffiths HR (2012) An aqueous extract of Fagonia cretica induces DNA damage, cell cycle arrest and apoptosis in breast cancer cells via FOXO3a and p53 expression. PLoS One 7(6):e40152

12. Turnell AS, Grand RJ (2012) DNA viruses and the cellular DNA-damage response. J Gen Virol 93(Pt 10):2076–2097

13. Hollingworth R, Grand RJ (2015) Modulation of DNA damage and repair pathways by human tumour viruses. Viruses 7(5):2542–2591

14. Sirbu BM, Cortez D (2013) DNA damage response: three levels of DNA repair regulation. Cold Spring Harb Perspect Biol 5(8):a012724

15. Surova O, Zhivotovsky B (2013) Various modes of cell death induced by DNA damage. Oncogene 32(33):3789–3797

16. Hakem R (2008) DNA-damage repair; the good, the bad, and the ugly. EMBO J 27(4):589–605

17. Russell P, Nurse P (1986) cdc25+ functions as an inducer in the mitotic control of fission yeast. Cell 45(1):145–153

18. Molz L, Booher R, Young P, Beach D (1989) cdc2 and the regulation of mitosis: six interacting mcs genes. Genetics 122(4):773–782

19. Ayscough K, Hayles J, MacNeill SA, Nurse P (1992) Cold-sensitive mutants of p34cdc2 that suppress a mitotic catastrophe phenotype in fission yeast. Mol Gen Genet 232(3):344–350

20. Nurse P (1990) Universal control mechanism regulating onset of M-phase. Nature 344(6266):503–508

21. Castedo M, Perfettini JL, Roumier T, Andreau K, Medema R, Kroemer G (2004) Cell death by mitotic catastrophe: a molecular definition. Oncogene 23(16):2825–2837

22. Castedo M, Perfettini JL, Roumier T, Valent A, Raslova H, Yakushijin K et al (2004) Mitotic catastrophe constitutes a special case of apoptosis whose suppression entails aneuploidy. Oncogene 23(25):4362–4370

23. Vitale I, Galluzzi L, Castedo M, Kroemer G (2011) Mitotic catastrophe: a mechanism for avoiding genomic instability. Nat Rev Mol Cell Biol 12(6):385–392

24. Galluzzi L, Vitale I, Abrams JM, Alnemri ES, Baehrecke EH, Blagosklonny MV et al (2012) Molecular definitions of cell death subroutines: recommendations of the nomenclature

committee on cell death 2012. Cell Death Differ 19(1):107–120
25. Vakifahmetoglu H, Olsson M, Zhivotovsky B (2008) Death through a tragedy: mitotic catastrophe. Cell Death Differ 15(7):1153–1162
26. Vitale I, Galluzzi L, Senovilla L, Criollo A, Jemaa M, Castedo M et al (2011) Illicit survival of cancer cells during polyploidization and depolyploidization. Cell Death Differ 18(9):1403–1413
27. Puck TT, Marcus PI (1956) Action of X-rays on mammalian cells. J Exp Med 103(5):653
28. Rafehi H, Orlowski C, Georgiadis GT, Ververis K, El-Osta A, Karagiannis TC (2011) Clonogenic assay: adherent cells. J Vis Exp 49:2573
29. Munshi A, Hobbs M, Meyn RE (2005) Clonogenic cell survival assay. Methods Mol Med 110:21–28
30. Castedo M, Perfettini JL, Roumier T, Yakushijin K, Horne D, Medema R et al (2004) The cell cycle checkpoint kinase Chk2 is a negative regulator of mitotic catastrophe. Oncogene 23(25):4353–4361
31. Rello-Varona S, Kepp O, Vitale I, Michaud M, Senovilla L, Jemaa M et al (2010) An automated fluorescence videomicroscopy assay for the detection of mitotic catastrophe. Cell Death Dis 1:e25
32. Guzman C, Bagga M, Kaur A, Westermarck J, Abankwa D (2014) ColonyArea: an ImageJ plugin to automatically quantify colony formation in clonogenic assays. PLoS One 9(3):e92444
33. Rello-Varona S, Vitale I, Kepp O, Senovilla L, Jemaa M, Metivier D et al (2009) Preferential killing of tetraploid tumor cells by targeting the mitotic kinesin Eg5. Cell Cycle 8(7):1030–1035
34. Blajeski AL, Phan VA, Kottke TJ, Kaufmann SH (2002) G(1) and G(2) cell-cycle arrest following microtubule depolymerization in human breast cancer cells. J Clin Invest 110(1):91–99
35. Khongkow P, Gomes AR, Gong C, Man EP, Tsang JW, Zhao F et al (2016) Paclitaxel targets FOXM1 to regulate KIF20A in mitotic catastrophe and breast cancer paclitaxel resistance. Oncogene 35(8):990–1002
36. Michalakis J, Georgatos SD, Romanos J, Koutala H, Georgoulias V, Tsiftsis D et al (2005) Micromolar taxol, with or without hyperthermia, induces mitotic catastrophe and cell necrosis in HeLa cells. Cancer Chemother Pharmacol 56(6):615–622
37. Ruegg UT, Burgess GM (1989) Staurosporine, K-252 and UCN-01: potent but nonspecific inhibitors of protein kinases. Trends Pharmacol Sci 10(6):218–220
38. Feoktistova M, Geserick P, Leverkus M (2016) Crystal violet assay for determining viability of cultured cells. Cold Spring Harb Protoc 2016(4):pdb prot087379
39. Senovilla L, Vitale I, Martins I, Tailler M, Pailleret C, Michaud M et al (2012) An immunosurveillance mechanism controls cancer cell ploidy. Science 337(6102):1678–1684
40. Bloy N, Sauvat A, Chaba K, Buque A, Humeau J, Bravo-San Pedro JM et al (2015) Morphometric analysis of immunoselection against hyperploid cancer cells. Oncotarget 6(38):41204–41215
41. Senovilla L, Demont Y, Humeau J, Bloy N, Kroemer G (2017) Image Cytofluorometry for the quantification of ploidy and endoplasmic reticulum stress in cancer cells. Methods Mol Biol 1524:53–64

INDEX

A

Apoptosis 8, 140, 160, 181–189, 207, 218, 228, 236

B

5-Bromo-2'-deoxyuridine (BrdU)................58, 64, 148, 151, 154, 155, 166, 169, 173, 174, 178

C

Calreticulin 208–212
Cell-cell fusion.. 145–155
Cell cycle......................................1, 8, 19, 70, 82, 83, 91, 103–106, 116, 125, 126, 128–139, 145, 146, 150, 151, 160, 165–178, 181, 182, 191, 192, 207, 218, 227, 228, 236, 237
Cell cycle progression 145–155, 165, 191, 192
Cell cycle regulation..................................... 145, 160, 192
Centrin... 231, 234, 237
CHK1 .... 1–6, 128, 129, 133, 135, 137, 191, 192
Chromatin 2, 21, 27, 43, 44, 52, 81, 82, 104, 106, 107, 110–117, 121–129, 131, 132, 134, 137, 138, 140, 152, 154, 192, 201–203, 237
Chromatin immunoprecipitation sequencing (ChIP-seq) 20, 21, 27, 28, 41, 43–45, 49, 52
Clonogenic assays... 228–232
Cytometry 58, 150, 165–178, 209–214

D

DNA affinity purifications 82, 83, 86–89
DNA-binding protein complexes........................ 8, 21, 81
DNA damage.......................... 7, 13, 14, 19, 57, 58, 103, 111–113, 125, 127, 129–135, 137, 140, 181–183, 185, 191, 192, 196, 218, 227, 228
DNA damage response (DDR)............................ 1, 2, 57, 103–141, 227
DNA fiber spreading assay..57–71
DNA-protein binding 83, 86, 87, 89
DNA repair 1, 103, 106, 125, 129, 181, 192, 196, 214, 225, 228, 237
DNA replication 57, 58, 104–106, 113–117, 123, 124, 126–131, 133, 135, 146
Double-stranded DNA brake (DSB) 111, 112, 128, 131, 132, 135, 192, 195, 197–200, 228
Doxorubicin 13, 14, 16, 162, 184, 185

E

E2 Cdc34 .... 91–96, 98
eIF2α..208, 217–225
eIF4E 74, 75
Endoplasmic reticulum (ER)..................... 208, 217, 218, 223, 224
End resection... 195
5-Ethynyl-2'-deoxyuridine (EdU) 148, 166
Eukaryotic translation initiation factor (eIF2α) 208, 217

F

Flow cytometry 58, 150, 165–178, 183, 209
Fluorescence resonance energy transfer (FRET)..91–102
Fluorescently labeled ubiquitin 74, 97–98, 100, 101
Functional complementation 145–155

G

Gene expression 19, 20, 73, 81, 104, 159
Gene knockdowns.. 161
Gene silencing 159
Global run-on sequencing (GRO-seq) 20–27, 31–41, 49, 51–54

H

HCl extraction 6, 155
Heterokaryons........................... 147, 150, 151, 153–155
Histones................................ 10, 13, 14, 27, 28, 49, 81, 83, 113, 114, 133–136, 229
Hoechst 33342 .... 149, 166, 168, 169, 172, 173, 175–178, 211, 220, 222, 223

I

Immunofluorescence 2, 58–60, 64, 65, 70, 146, 149, 151, 153, 196, 197, 208, 209
Immunofluorescence microscopy 1–6, 217–225
Immunogenic cell death (ICD) 208, 218

K

Ki-67 .... 165–168, 170–172, 177
K48 ubiquitination 91, 93, 99

James J. Manfredi (ed.), *Cell Cycle Checkpoints: Methods and Protocols*, Methods in Molecular Biology, vol. 2267, https://doi.org/10.1007/978-1-0716-1217-0, © The Author(s) 2021

Index

M
Mass spectrometry 81, 82, 87, 89
Mitotic catastrophe 207–214, 217–225, 227–237

N
Nutlin 22, 41, 48, 49, 53

O
Origin firing 58, 68, 128

P
p53 .... 7–17, 19–54, 58, 75, 82, 84, 104, 128, 159, 160, 162, 181, 182, 191
Phosphor-specific antibodies 15, 133
Phosphorylation 1, 2, 7, 8, 13–16, 113–115, 128, 129, 133–138, 182, 192, 208, 217–225
Phylogenetic footprinting 83
Post-translational modifications 7, 8, 13, 129
Promoters elements 83–85
Propidium iodide (PI) 166–169, 171, 172, 174, 176, 178, 201, 202, 213
Protein shuttling 147
Pyronin Y (PY) 166, 168, 169, 172, 173, 176–178

Q
Quantitative PCR (qPCR) 44, 53, 194, 199, 200, 203

R
Replication fork stalling 58, 66, 68
RNA immunoprecipitation (RNA IP) 73–78
RNA interference (RNAi) 159
RNA sequencing (RNA-seq) 20, 21, 29, 46–49, 73

S
Schizosaccharomyces pombe 192, 197, 228
Sequencing 21, 25–30, 33–41, 44, 45, 47, 48, 50, 53, 74
Short interfering RNA (siRNA) 104, 159–163
S-phase 57, 105, 106, 118, 121, 125, 128, 129, 131, 133, 146, 148, 150, 152, 154, 165, 166, 182
Stable isotope labeling with amino acids in cell culture (SILAC) 82

T
Taxanes 208, 211, 218, 228
Transcription 21, 37, 41, 48, 52, 53, 57, 73, 75, 84, 104, 126, 181
Transcription factor binding sites 83
Transcription factors 7, 19, 21, 81–83, 87, 147, 160, 218
Transcriptome 21, 46–48, 54
Translation 73–79, 218
TUNEL assays 181–188

V
Vinca alkaloids 208, 211, 218

W
Western blot 9, 11, 12, 14, 82, 83, 87, 89, 161, 162, 202

X
Xenopus cell-free extracts 104, 126, 134